Electronic Communications with Arduino

A Project Based Approach

By Bob Dukish

Copyright 2022 by Dukish LLC

All rights reserved. This book may not be reproduced and may not be copied by any means, including electronic, mechanical, recording or photocopy, or by any document storage and retrieval system, without the written permission of the author. The projects are presented without any claim of suitability.

Website: www.dukish.com
For information: bob@dukish.com
Published by Dukish LLC
ISBN: 9798843872083

Parts information and code downloads are available at www.dukish.com

Warning:
Electrical circuits and components may contain lethal voltages even when disconnected. Always refer to the parts data sheets. Do not attempt to test, modify, or repair electrical equipment. Hazardous voltages are present when equipment covers are removed.

Preface

The fantastic electronic technology that drives our world comes from the need for humans to communicate over long distances. It evolved from audible and visual signals, to the control of electricity to produce wireless messages. The results of our inventiveness have been exponential. The first worldwide commercial radio broadcast was in 1920. In less than a century, the spacecraft Voyager 2 left the solar system in 2013 with a golden record containing the sounds of our planet. There have been tremendous advancements in science and technology in the last decade as we now create artificial intelligence. When we stop and think of how ubiquitously the world is now interconnected, we see how electronic communications affects our everyday life.

This book features many "hands-on" projects with modern communications devices as we present the fundamental science behind wireless connectivity. We keep the mathematics to a minimum, with the material presented in a well-paced, understandable way. The Arduino microcontroller and associated components aid simulations and provide a low-cost method of constructing operational projects.

Acknowledgments

I would like to thank my students. I learned a great deal from them over the years. I also thank the broadcast engineers I worked with over the years and appreciate their passion for adhering to best practices.

About the author

Bob Dukish has spent over 40 years working and teaching in the field of technology. After serving in the military, working as an electronics component engineer, and running a corporation, he now teaches at Kent State University. He has Associate Degrees in Avionic Systems, and Electronics Engineering Technology. A Bachelor's Degree in Physics from Syracuse University, as well as Master's degrees from Kent State University, and Rensselaer Polytechnic Institute. He earned his last degree at age 54 and considers himself a lifelong learner.

Table of contents

Chapter 1. A brief look at electromagnetism.
    Section 1.1. Forces acting at a distance   7
    Section 1.2. Types of electromagnetic waves   9
    Section 1.3. Radio frequencies   12
    Section 1.4. Audio frequencies   15
    Section 1.5. Audio frequency project   17
    Section 1.6. Audio amplifier project   21
    Section 1.7. Bandwidth   24

Chapter 2. Infrared communication and lasers
    Section 2.1. IR devices   27
    Section 2.2. IR advantages and limitations   29
    Section 2.3. Laser and fiber optics   30
    Section 2.4. IR light reflection project   32
    Section 2.5. IR direct line-of-sight project   35
    Section 2.6. IR control link project   36

Chapter 3. Bluetooth, Wi-Fi, and ISM devices
    Section 3.1. Bluetooth and BLE   43
    Section 3.2. Wi-Fi   45
    Section 3.3. ISM devices   47
    Section 3.4. Bluetooth remote temperature sensor project   48
    Section 3.5. Bluetooth controlled lighting project   54
    Section 3.6. 433 MHz transmitter and receiver project   63

Chapter 4. Wired and wireless communications
    Section 4.1. The telegraph and CW   69
    Section 4.2. Square to sine wave project   73
    Section 4.3. Resonance and oscillators   74
    Section 4.4. CW transmitters and receivers   77
    Section 4.5. I2C communication project   79
    Section 4.6. Morse code project   82

Chapter 5. Noise
- Section 5.1. Electron movement in conductors — 85
- Section 5.2. Electron movement in semiconductors — 86
- Section 5.3. External noise — 87
- Section 5.4. Noise specifications — 89

Chapter 6. Broadcast communications
- Section 6.1. Amplitude modulation — 95
- Section 6.2. Classes of amplifiers — 100
- Section 6.3. Pulse width modulation project — 102
- Section 6.4. Frequency modulation — 103
- Section 6.5. Television — 106

Chapter 7. Broadcast receivers
- Section 7.1. AM receivers — 111
- Section 7.2. FM receivers — 115

Chapter 8. Antennas
- Section 8.1. Vertical antennas — 119
- Section 8.2. Horizontal antennas — 122

Appendix
- Section A.1. Transducers — 127
- Section A.2. Amplification — 128
- Section A.3. Operational amplifiers — 131
- Section A.4. Decibel information — 133
- Section A.5. Parts list — 134

Notes:

Hook-up wires are recommended to replace switches in projects.

Appendix sections A.1 through A.3 contains excerpts from the book Extreme Fundamentals of Technology by Bob Dukish.

Chapter 1
A brief look at electromagnetism

Section 1.1. Forces acting at a distance

I find mechanical forces more straightforward to analyze than invisible forces like radiation and gravity. We know the correct set of barbells in the gym to give us a good workout. But it's sometimes difficult to judge the time of exposure to summer sun before receiving a burn or respecting the high temperature of a propane gas grill. It's early summer as I'm writing this book, and I have red arms and burned fingers. As we work through the projects in the book, we must seriously respect the force of electricity since it can be dangerous when not handled properly. Even at low voltages, avoid inadvertently touching conductors across voltage potentials, producing a short circuit.

The universe has many derivations of the four fundamental forces: The strong nuclear force (The force within the nucleus of atoms), the weak nuclear force (radioactive decay), gravity, and our subject of interest - electromagnetism. The forces of electricity and magnetism have reciprocity and can act at a distance. When you pass an electric current through a conductor, corresponding magnetic lines called flux will circulate around and perpendicularly to the conductor. On the other hand, electricity is produced if there is movement between a magnetic field and a conductor.

In a wired loop, the current can light a lamp or produce some other type of work. In looking at a straight conductive wire, we can use the left-hand rule to describe the direction of the magnetic flux lines, as shown in Figure 1.1.

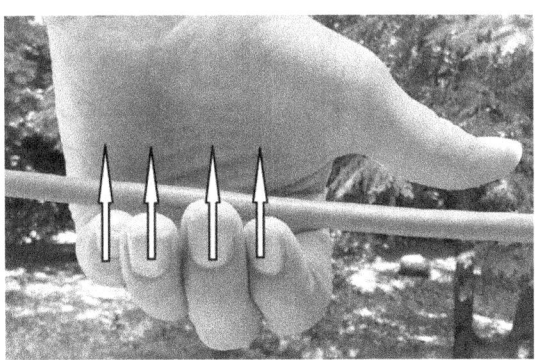

Figure 1.1. The Left-Hand Rule

With the thumb pointing in the direction of electron current flow, the magnetic lines of flux form concentric circles perpendicular to the wire, where their path is upward and matches the direction of the fingers shown in front of the wire. The case described is for DC. In the case of AC, the directions would alternate and increase and decrease in intensity in a sinusoidal manner.

The interaction between electric current flow and magnetism allows for the existence of life on our planet. There is a large core of molten iron inside the earth with a tremendously powerful circular current flow. The current flow produces the magnetic field above the earth's surface called the magnetosphere. When the sun occasionally ejects particles, in what is termed a Coronal Mass Ejection (CME), we are treated to a northern light show called the Aurora Borealis. In the southern hemisphere, the light show is called the Aurora Australis. The shimmering lights are not only beautiful to observe, but the high-energy particles' interaction with the earth's magnetic field also keeps the high-energy at the edge of the atmosphere, at the top of the stratosphere. Average solar radiation produces a charged area in this region as well. Its name is the ionosphere. The charged area of the ionosphere allows communications using short-wave radio signals from just below 3, up to 30 MegaHertz to reflect between, sometimes very distant points across the globe. However, a CME may also produce an immediate x-ray pulse which can wipe out the reflective property of the ionosphere for a short time. The x-ray energy pulse hits us in about 8 minutes, whereas the ejected particles may take a few days to reach the earth. Without the magnetosphere around our planet, the atmosphere would be wiped away. As we continue to explore the Moon and Mars, solar and other cosmic radiation is considered a risk factor. There is no magnetosphere around the Moon, and there is no atmosphere. Mars has a feeble magnetic field, and consequently, the atmosphere is relatively thin. Without a robust magnetosphere for protection, a planet's surface is more susceptible to bombardment from dangerous high-energy particles, and life cannot exist without some type of shielding.

We joked about being in the sun for too long of a time and receiving a sunburn. We also mentioned radiation. Usually, people associate the word nuclear with radiation, but just as in our left-hand rule discussion, electromagnetism is radiation of electricity and magnetism that can produce forces at a distance. The radiant energy from the sun is an example of electromagnetic radiation and has the components of perpendicular electric and magnetic fields, as illustrated in Figure 1.2.

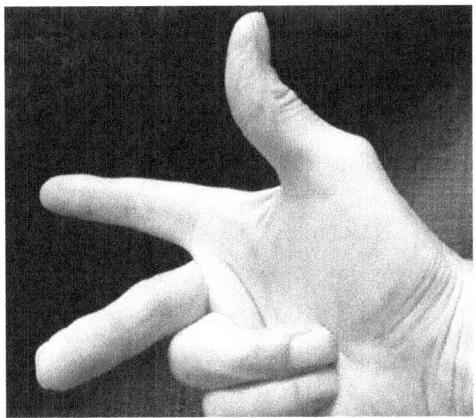
Figure 1.2. A hand illustrating a transverse wave

Even though it appears the subject in the figure is flashing a gang sign, orienting one's hand in this way conveniently represents the transverse nature of an electromagnetic wave. If the forefinger in the picture is pointing in the direction of the electromagnetic propagation, the thumb represents the magnetic field, and the middle finger represents the electric field. The fields alternate in a sinusoidal manner as the wave travels through space at the speed of light.

Section 1.2. Types of electromagnetic waves

There is a saying to describe the entire electromagnetic spectrum as containing everything from DC to daylight. Every free-traveling AC wave indeed has a place on the spectrum. While many characteristics are similar, there are some peculiarities at different frequencies. For convenience, we generally describe waves with lower alternations in terms of frequency and those at faster alternations in terms of wavelength. The relationship between wavelength and frequency is inverse and has the equation: $\lambda = \frac{c}{f}$. Where λ is the Greek lowercase symbol Lambda used to represent the quantity of wavelength, The letter c represents the speed of light equal to 186,282.397 miles/sec, in free space, which is 983,571,056 feet/sec, divide by 3 for yards - if you wish. It is much cleaner and easier to remember using the metric system, which is 299,792,458 meters/sec. (To make life even more straightforward, in this text, we round off the speed of light to 186,000 miles/sec, or 300,000,000 meters/sec.)

Problem

Find the wavelength λ in meters of a 5G cell phone operating on a frequency of 7.125 GHz.

Solution

Keeping in mind the Engineering Prefixes are kilo 1×10^3, Mega 1×10^6, Giga 1×10^9, and Tera 1×10^{12}, our problem gives us a frequency of 7.125×10^9. The formula $\lambda = \dfrac{c}{f}$ is directly used giving:

$$\lambda = \frac{300^6}{7.125^9} = 0.042 \text{ meters, or } 4.2 \text{ cm.}$$

Besides cooking food, microwaves are mainly used for line-of-sight radio frequency (RF) purposes. While there are additional subdivisions, the usable RF spectrum is generally shown in Chart 1-1.

Low	Medium	High	VHF	UHF	Microwave
3Hz to 300 kHz	300 kHz to 3 MHz	3 MHz to 30 MHz	30 MHz to 300 MHz	300 MHz to 3 GHz	Above 3 GHz

Chart 1-1. The RF spectrum

Electromagnetic waves, much shorter than microwaves, take on a noticeable duality of both wave structure and some characteristics of discrete particles and are called photons. We can best explain the duality by considering that the energy becomes more packetized at higher frequencies. Planck's constant comes in handy and is approximately 6.63×10^{-34} Joules/sec and has the symbol h. (Energy units are called Joules and are similar to volts.) The formula $E = hf$ gives the energy of a photon and mathematically describes the quantum mechanical nature of the massless, extremely short-wavelength energy packet. Chart 1-2 identifies the descriptions of the energy as the wavelengths become shorter. Each chart section encompasses a band (group) of wavelengths with a representative value. (940 nm is the standard version of a commercial Infrared LED, which we will use in later projects.)

IR	Light	UV	XRAY	Gamma	Cosmic
1 micro meter (um)	500 nano meter (nm)	100 nano meter (nm)	1 nano meter (nm)	1 pico meter (pm)	Shorter than 1 pm

Chart 1-2. Descriptions of shorting wavelengths

Problem

Compare the energy packet of a lower frequency to one much higher. We will use the wavelength of 1 um from the IR band of the chart and compare it with a 100 nm wavelength from the UV band.

Solution

To use the formula given earlier for the energy, we first must change the given wavelengths to frequencies.

For IR: $\lambda = \frac{300^6}{f}$, and $f = \frac{300^6}{\lambda} = f = \frac{300^6}{1^{-6}} = 3^{14} = 300$ TeraHertz (THz)

For UV: $f = \frac{300^6}{\lambda} = f = \frac{300^6}{100^{-9}} = 3^{15} = 3000$ THz

Since we now have both frequencies, we can use the given energy formula, where Planck's constant is approximately 6.63×10^{-34} Joules/sec.

For IR: $E = hf = (6.63^{-34})(3^{14}) = 2^{-19}$
For UV: $E = hf = (6.63^{-34})(3^{15}) = 2^{-18}$

This problem demonstrates that as electromagnetic energy goes higher in frequency and shorter in wavelength, the signal's energy goes up. This is an important concept to remember! It is more hazardous when working with higher frequencies. (i.e., microwave transmissions are generally more dangerous than CB or HAM radio transmissions.) Care must also be taken when electromagnetic energy is passively amplified by focus creating directionality of the signal.

Section 1.3. Radio frequencies

We earlier mentioned that great distances could be covered by the ability of high-frequency (HF) short-wave (SW) radio waves to reflect back down to earth from the ionosphere, located above the atmosphere and stratosphere. A few different reflective layers form the ionosphere, as depicted in Figure 1.3.

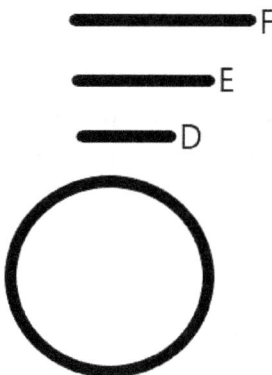

Figure 1.3. Ionosphere layers

During the daytime, the D layer exists at about 30 miles above the earth, and the F layer splits into two sections with the highest altitude at roughly 300 miles. At night, the F layer combines, and the D layer disappears. The layers' distances are quite variable depending on the amount of solar radiation the earth receives and is also dependent on Sunspots. More Sunspot activity usually provides greater ionization and aids in the reflection of radio signals. Reflection also is dependent on the radio frequencies used and the angle of transmission towards the ionosphere.

Very low radio frequencies do not usually travel up towards the ionosphere, and any that make it are absorbed. Very low frequencies tend to follow the curvature of the earth. With a powerful transmitter, they can travel quite far. Highly accurate home consumer clocks, sometimes called Atomic Clocks, are synchronized with the U.S. National Institute of Standards and Technology (NIST). They operate a powerful transmitter in Colorado synchronized with an atomic clock. The transmitter's frequency is 60 KiloHertz (kHz) and covers most of North America. Lower frequencies are used for submarine communications and are given the designation of Extremely Low Frequencies (ELF). They are used because they follow the curvature of the earth as well, but are also not as significantly attenuated by water. Higher frequencies tend to be more line of sight and may be blocked by obstacles.

The radio frequency spectrum's major classifications appear in Chart 1-3.

Extremely /Low (ELF/LF)	Medium (MF)	High (HF)	Very High (VHF)	Ultra High (UHF)	Super/ Extremely High (SHF/EHF)
3 kHz to 300 kHz	300 kHz to 3 MHz	3 MHz to 30 MHz	30 MHz to 300 MHz	300 MHz to 3 GHz	3 GHz to 300 GHz

Chart 1-3. The RF frequency ranges.

The International Telecommunications Union (ITU) is responsible for allocating specific frequencies for different services. Each member nation also can determine particular usage of the spectrum. In the United States, the Federal Communication Commission (FCC) designates allocations and is responsible for enforcing the rules and regulations. Space on the radio frequency portion of the electromagnetic spectrum is quite limited and therefore very valuable. Very few sections are unlicensed. However, small segments of unlicensed radio spectrum are used for Industrial, Scientific, and Medical (ISM) equipment, where radio frequency signals may radiate as a byproduct of device operation. A typical example is the microwave oven which operates at 2.45 GHz. Limitations on radio frequency emissions in the unlicensed ISM bands are usually kept to a minimum and may not exceed specified values. The value for a microwave oven is 5 milliwatts/cm^2 at a 5 cm distance from the unit's case. Microwave oven energy leakage will usually occur around the door seal and is checked with a device similar to the leakage tester shown in Figure 1.4.

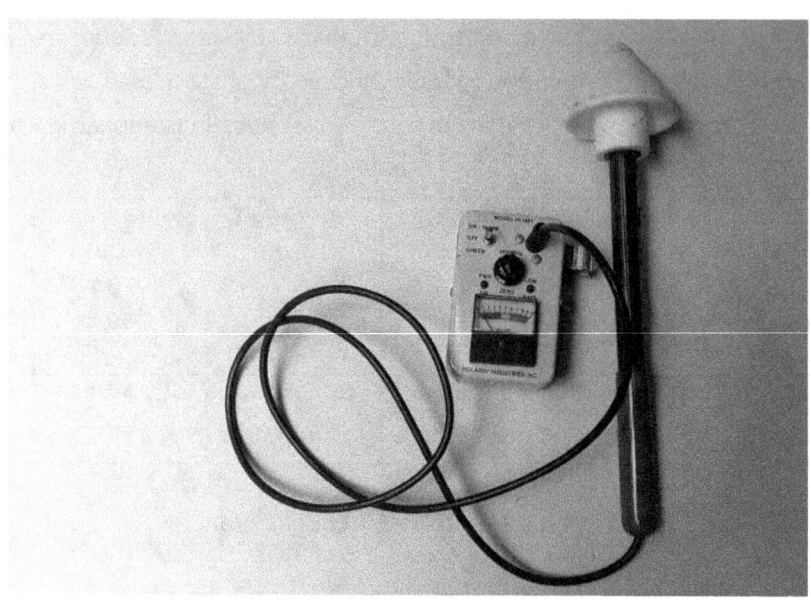

Figure 1.4. Microwave oven leakage detector

The device is a sensitive receiver tuned for transmissions at 2.45 GHz and must be regularly calibrated. A cup of water is heated on the highest cooking level while the handheld wand is swept perpendicularly around the edge of the door seal. If a higher indication than the acceptable amount is detected, the microwave must be repaired or replaced. The testing is mainly related to the safety of the consumer but also meets the requirements for limited spurious electromagnetic radiation.

Regulated services sometimes share frequency allocations with unlicensed ISM equipment. An example is on some channels of the unlicensed Wi-Fi band where the 2.3 GHz Amateur Radio band is the primary user and is limited to 1500 watts of output power. In contrast, some unlicensed Wi-Fi channels in that range have a limitation of under 20 milliwatts. The Wi-Fi service is considered to be secondary and is subject to interference. The Amateur Radio service's best practice is to use the lowest amount of power possible to establish a communications link. Directional antennas can also be used to provide passive power gain and reduce interference.

Popular ISM transmitters and receivers are available to hobbyists who can operate in either the 433 MHz and 915 MHz ranges, depending on the ITU region. These devices are subject to interference and have a very short range. A pair of the 433 MHz devices are pictured in Figure 1.5.

Figure 1.5. 433 MHz transmitter and receiver.

These devices are used to send digital signals with less than 25 milliwatts of power. These types of devices use different frequencies and maximum power outputs depending on the part of the world (ITU region). When a transmission is detected, the receiver outputs a high pulse level which reduces power consumption. A program can check for the presence of a pulse, or the pulse width duration. More functional programs can recognize a coded pulse train to perform an activity. A pulse train is a series of pulses. The antenna length is approximately just under 7 inches for a quarter-wavelength at 433 MHz, but the bottom may be coiled to reduce the overall length. Both the transmitter and receiver units must have antennas.

Section 1.4. Audio frequencies

The human audio range is from 20 Hz to 20 kHz. People with excellent hearing may hear as high as 22 kHz, but the ear's response to high frequencies tends to lessen with age. A joke uses this to explain how marriages can last into the golden years. Acoustic frequencies are sometimes used for special purposes. Ships, submarines, and sonobuoys can passively detect submerged objects by receiving their sounds from underwater *transducers*. Transducers are devices that convert energy from one form to another. So, an underwater microphone can be used to detect sound from objects within a water environment. Multiple Sound Navigation (Sonar) receivers can determine location utilizing the process of triangulation. Since they are passive, they are nearly undetectable and used by military and research vessels. Active sonar is like radar in that it emits a pulse and listens for an echo to determine distance. The speed of sound is minuscule compared to the speed of light; at room temperature, sound waves travel at just under 800 miles/hour. Along with temperature, the speed of the wave is also dependent on ambient air pressure. The time taken for an echo to be returned can be used to determine the distance to an object. A horizontal and vertical sweeping motion can also be used to determine the azimuth and elevation when ranging a moveable object. An active return pulse sonar uses both a sound transmitter and receiver. The use of lower frequencies travels the best underwater. The reflection of sound pulses through the air is sometimes used as a safety feature for vehicles to avoid contact with objects. Sensors for collision avoidance systems may be visible on a vehicle's exterior.

Sounds below the range of human hearing are called subsonic or infrasonic, and those above are termed *ultrasonic*. Infrasonic waves can be mighty and are sometimes naturally created when large meteors impact the atmosphere or when there are volcanos or other large explosions. These powerful acoustic waves can reverberate back and forth across the globe. Ultrasonic sound, located above the range of human hearing, can be used for applications like vehicle backup collision avoidance due to low power requirements and greater resolution. Many ultrasonic sensors are available to hobbyists. One of the most popular is the ping sensor shown in Figure 1.6, which has a range of up to 12 feet.

Figure 1.6. Ping ultrasonic sensor

The ping sensor in the figure shows two transducers; one is a speaker, and the other is a microphone. Electronic circuitry on the back of the board provides output to one of the three pins shown; the other two provide power and ground. (Some Ping sensors have four pins and separate the output pulse and echo.) The sensor shown in the figure uses a frequency of 40 kHz, well above the range of human hearing. The devices are inexpensive, and many Arduino software sketches are available online. These devices are typically used at the front end of a mobile robot and can be swept back and forth for object avoidance.

Section 1.5. Audio frequency project

Our first project is to program the Arduino to produce audio waves of different frequencies. It is essential to realize that audio waves are entirely different than radio waves. An audio wave consists of vibrations of the air and cannot exist in a vacuum (despite the extraordinary sound effects in some outer space movies.) Radio waves are a part of the electromagnetic spectrum and do not need a medium; they can travel through empty space. When electromagnetic waves propagate through a medium, they slow down and may also be attenuated.

The process of carrying audio information on an RF wave is called *modulation.* The first section of an analog modulator usually contains an audio amplifier and is a project that we will cover later in this chapter. When using an RF channel to carry modulated information, it must then be demodulated after the signal has been transmitted and reaches the receiver. Modulation is the process of altering the RF carrier signal to transfer the intelligence between two points in space. Demodulation recovers the information from the altered carrier. On computer networks, modulation and demodulation are put together in the modem. We will cover many of the standard modulation techniques throughout subsequent chapters.

Our first project is the production of audio waves of differing frequencies. Both pure audio waves and electromagnetic waves are sinusoidal and appear as in Figure 1.6.

Figure 1.6. Representation of a sine wave

For a digital circuit to produce a sine wave, it must have a Digital to Analog Converter (DAC). The Arduino does not contain a DAC but produces sound by outputting a square wave, as pictured in Figure 1.7.

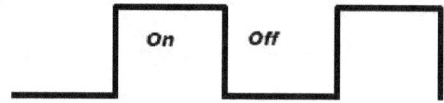

Figure 1.7. Representation of a square wave

You can program an Arduino to produce sound in two ways. The most straightforward method is to toggle a high to low pulse so that each cycle's time period is the inverse of the frequency to be generated. The coding is a bit tedious. The second method uses a built-in Arduino command called *tone* to do the toggling. We use the tone method in Code Listing 1.1. The Arduino coding program can be downloaded at no cost at arduino.cc. Links to code examples are also available from my Website dukish.com. The Arduino programming screen is called an Integrated Development Environment (IDE) and is shown in Figures 1.8.

```
sketch_jun30a | Arduino 1.8.15 (Windows Store 1.8.49.0)
File Edit Sketch Tools Help

sketch_jun30a
void setup() {
  // put your setup code here, to run once:

}

void loop() {
  // put your main code here, to run repeatedly:

}
```

Figure 1.8. Arduino IDE

Arduino programs are called *Sketches* and given the default name of the date when the code was written. Also, by default, the two main areas to include code are shown. Our program in Code Listing 1.1 will produce three different audio tones to a speaker connected to pin 7 through a 120 Ohm resistor. (it is vital to have the series resistor in place to limit current flow through the speaker, or damage may result.)

```
const int fiveHundred = 8;
const int oneThousand = 9;
const int fiveThousand = 10;
int first;
int second;
int third;

void setup() {
  pinMode (fiveHundred, INPUT_PULLUP);
  pinMode (oneThousand, INPUT_PULLUP);
  pinMode (fiveThousand, INPUT_PULLUP);
  pinMode (7, OUTPUT); //the sound output connects to pin 7
}

void loop() {
  first = digitalRead (fiveHundred);
  delay (20); //debounce
  second = digitalRead (oneThousand);
  delay (20);
  third = digitalRead (fiveThousand);
  delay (20);
  if (first == LOW) { //for 500 Hz
    tone (7, 500);
    delay (5000);
    noTone (7); //shuts off
  }
  else if (second == LOW) { //for 1 kHz
    tone (7, 1000);
    delay (5000);
    noTone (7); //shuts off
  }
  else if (third == LOW) { //for 5 kHz
    tone (7, 5000);
    delay (5000);
    noTone (7); //shuts off
```

}
}

Code Listing 1.1. Three tone output

After coding the sketch, you may need to select the Arduino communications port before uploading it to the board. Usually, it will be the highest number appearing in the port selection under the *tools* menu, as shown in Figure 1.9.

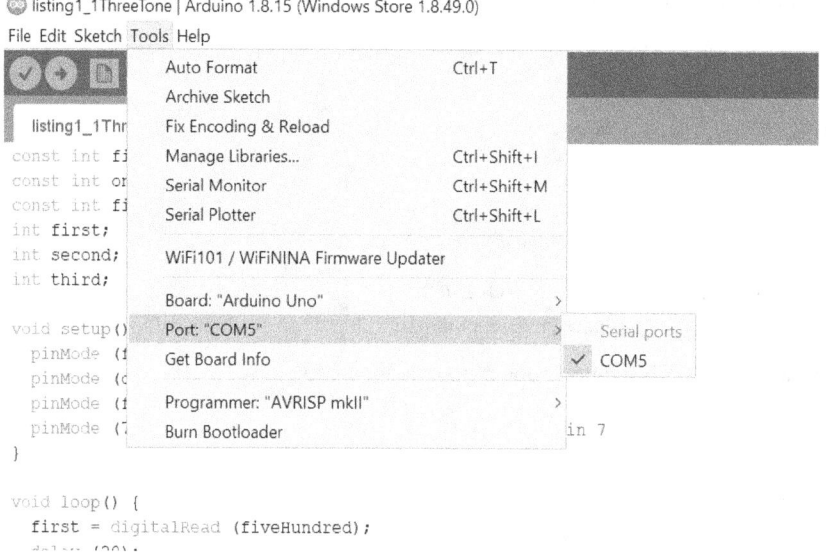

Figure 1.9. The port selection

The sketch in Code Listing 1.1 can be uploaded by clicking the arrow located next to the checkmark in the IDE. Once uploaded, momentarily tapping a small hook-up wire from the appropriate Arduino pin to a ground, such as the metal on the USB connector case, will cause a five-second tone from the speaker, connected between pin 7 and ground through the 120 Ohm resistor. See Figure 1.10.

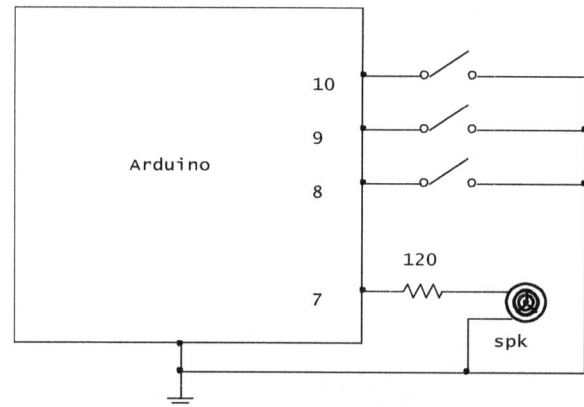

Figure 1.10. Three-tone project wiring diagram

In examining the code in Listing 1.1, we see that selection pins 8, 9, and 10 are given names. This process isn't necessary, but it's good practice and helps track the inputs and output functions. We skipped the naming process for the speaker output pin, number 7. We then defined the variables and assigned them as the integer type. To have saved Arduino memory space, we could have given them the Boolean type since they represent only High and Low levels. They are defined in the setup section as pulled-up inputs. This code saves us from providing three external resistors since this is accomplished internally in the ATmega 328P Arduino IC. A stand-alone pull-up circuit, however, is shown in Figure 1.11.

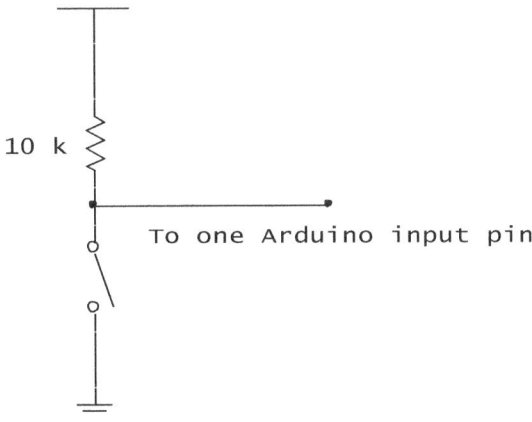

Figure 1.11. Digital pull-up circuit

The resistor limits current but provides a high digital level to the Arduino input pin when the switch is not grounded. There would be a short to ground without the resistor in place.

In the main loop of the program, each input logic level is examined, and the *if* and *else if* conditional statements will provide a tone to the speaker if one is found to be a low logic level when a wire (acting as a switch) is momentarily connected to ground by tapping the USB shield box.

Section 1.6. Audio amplifier project

Audio amplifiers are used in both transmitters and receivers dealing with broadcast signals like radio and television and two-way communication. The modulation may be digital, but voice, music, and video are analog. We will now investigate an audio circuit without RF in this project. The circuit uses a very common audio power amplifier IC found in many consumer devices such as amplified computer speakers, small televisions, and other A/V devices. The main IC is an 8-pin chip with the part number LM386. The pin numbering system is shown in Figure 1.12.

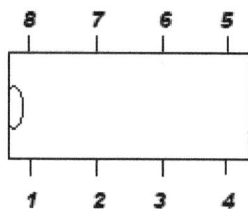

Figure 1.12. Pin numbering of an IC

Integrated Circuits (ICs) with pins located on both sides, like the one in the figure are called Dual Inline Packages (DIPs). They will have a semicircle or orientation mark denoting the schema of the numbering system. The part number will also appear right-side-up. The pin located under the mark and to the left is assigned as pin 1. No matter how many pins there are in a DIP, they will have the pin numbers increase as you proceed across the bottom edge from left-to-right, and then from right-to-left across the top of the IC. A specific IC datasheet is needed to determine the pin functions (called the pin-out).

In our project, we will build the analog circuit as shown in the schematic Figure 1.13. The capacitor values are given in microfarad (uF.) The large capacitor connected between the output and speaker passes the sound but blocks DC. The LM386 is designed to drive a speaker, and no current limiting resistor is needed.

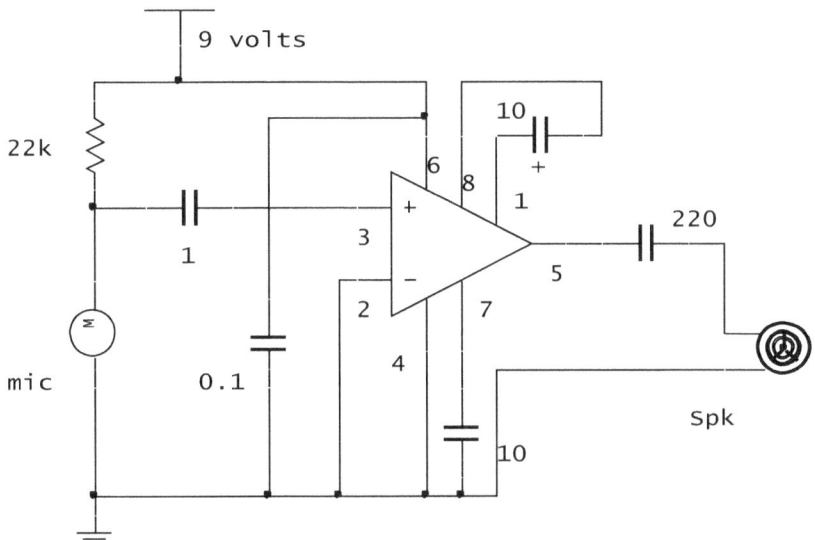

Figure 1.13. Microphone Amplifier

In the configuration shown in the schematic, a gain of 200 is achieved. The coupling across pins 1 and 8 determines the gain. The gain is a power multiplication factor. The microphone we use is called an electret condenser and works with a pull-up resistor to provide voltage to operate a small amplifier within the microphone. The metal microphone case is grounded through one of the bottom leads. The breadboarded circuit is shown in Figure 1.14.

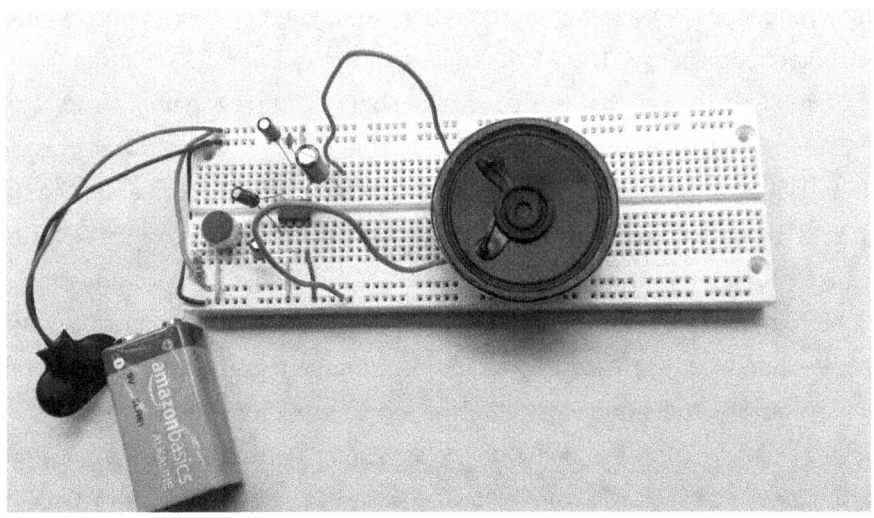

Figure 1.14. Breadboard circuit of an audio amp

In the orientation shown in the figure, the individual horizontal rows on both edges of the breadboard are connected and usually are used as power busses. It is standardized that each row across the edges is connected; however, there is a break in the middle of each row on some breadboard models. This should be tested before the board is populated with parts. If there is a break from the left to the right side, jumper wires can easily be inserted to connect the entire set of holes.

The LM386 power amplifier is able to operate anywhere between 5 and 12 volts. We are using a 9-volt transistor battery. The IC is not really designed to directly amplify the low power of a discrete microphone. There should be a preamplifier located between the microphone element and the 386 power amp. We are skipping the preamplifier in this project to simply demonstrate audio amplification. When the battery is connected and a person speaks directly into the microphone, the audio should output from the speaker. Due to our simplification of the circuit, the quality will not be optimal. If there is a feedback howl, you may try moving the speaker slightly away from the microphone.

Section 1.7. Bandwidth

Audio waves are mechanical vibrations and travel through the air or through materials. Despite great special effects in science fiction movies about outer space, when the phasors fire and an enemy ship blows up, there is no sound to be heard. Radio Frequency (RF) energy, on the other hand, can be thought of as vibrations of electromagnetic energy. The RF range was shown in Chart 1-3 and ran from about 3 kHz to 300 GHz. Above that are IR, visible light, UV, X-ray, gamma and cosmic rays. You can equate useful RF sections with valuable frontage along a well-traveled commercial highway where space is at a premium. Just as commercial lot sizes are kept to a minimum to conserve cash outlays, RF bandwidth is conserved. Channel sizes are regulated in the U.S. by the Federal Communications Commission (FCC) and internationally by the International Telecommunications Union (ITU). As we will later see, bandwidth is the group of adjacent frequencies needed to communicate, and high-quality audio and video information can require wide bandwidths. A center frequency usually contains maximum power, and side frequencies provide information. The range will usually form a bell-shaped curve, as drawn in Figure 1.15. Wider bandwidths are lower and broader, whereas High quality (Q) bandwidths are higher but narrower.

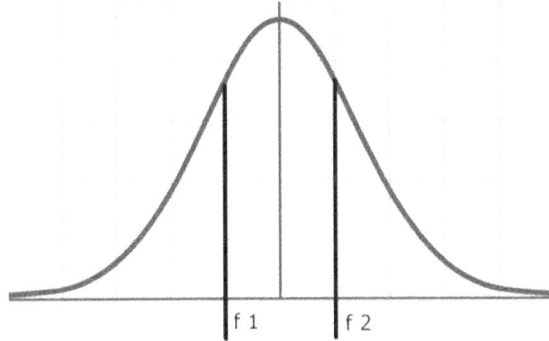

Figure 1.15. Bandwidth diagram

We are representing voltage on the y axis with reference to frequency on the x axis and f1 is lower in frequency than f2. The points labeled F1 and F2 are considered the cutoff points of the emissions. The frequency difference between the points is called the bandwidth. (BW = f2 – f1.) The cutoff points are found where the maximum voltage is 0.707 (70 %) at both above and below the center frequency. They are also the half power points. Some signal is emitted beyond the cutoff points, but it is generally ignored.

Chapter One Summary

Saying the electromagnetic spectrum contains everything from DC to Daylight is a simplification since our chart shows the high energy wavelengths above the visible light range consist of ultraviolet light, X-rays, gamma, and cosmic rays. Audio waves are acoustic and require a medium such as air to produce vibrations. Electromagnetic energy, however, needs no medium and can travel through empty space. The speed of light is the fastest anything can travel in the universe. The speed of sound is minuscule in comparison and depends on ambient temperature and air density. Radar uses radio waves to detect direction and distance, while infrasonic and ultrasonic devices use sound waves. Specific wavelengths of both light and sound behave differently. Long-distance radio communication is dependent on the charged area located above the atmosphere and stratosphere called the ionosphere. The amount of charging in that region is dependent on solar activity. The bandwidth of RF transmissions is the range of frequencies necessary to covey information.

Chapter One Questions

1. What are the four forces that seem to make up the universe?
2. Explain the reciprocity between electricity and magnetism.
3. What is the name of a planet's shield which protects it from high-energy particles?
4. What general frequencies hug the curvature of the earth as they travel?
5. Name a form of energy above radio frequencies.
6. In audio amplification, why is a capacitor used in series with a speaker?
7. Which travels faster, acoustic or electromagnetic energy?
8. What is the pin number located just under an orientation mark on a DIP.
9. The term DIP stands for what three terms?
10. Describe the term modulation.

Chapter 2
Infrared Communication and lasers

Section 2.1. IR devices

You probably have many IR communications devices in your home used for digital control of entertainment devices. The most popular are television remote controls. While some longer-distance remote controls use RF energy in the UHF band or sometimes Bluetooth Low Energy (BLE), most standard remote controls use infrared as the carrier medium. Remote controls that utilize IR are blocked by walls and other opaque objects, whereas RF and BLE signals may pass through things, albeit slightly attenuated in power. IR systems are line-of-sight but can slightly be reflected. IR is the preferred method for home entertainment unit remote control because of the simplicity and low cost, but they are limited in range up to about twenty feet. The general operation of both IR and RF remote control systems is that one device is a transmitter sending modulated coded data pulse trains to a receiver located in the television or other appliance. A microcontroller located within the receiver interprets the code and acts accordingly. One of our chapters' projects will be to construct an IR transmitter and receiver system to activate a series of sounds dependent on the user's selection.

Remote controls for home devices use a standard IR wavelength of 940 nanometers flashing at a square wave frequency of 38 kHz to produce carrier pulses. Most, but not all, equipment manufacturers adhere to this standard. Infrared radiation is located just below the visible light region of the electromagnetic spectrum. It is longer in wavelength and lower in frequency. Although not directly visible to the eye, CCD cameras can show the pulses from a remote control. As a troubleshooting technique, you can record a short video with a remotes' LEDs directed towards a camera as you push the buttons to ascertain the remotes' condition. Some replacement IR LEDs come in a matched set containing both transmitter and receiver diodes. Remote receivers more commonly come as an amplified unit, usually having three pins to connect to a circuit board. The pins are for voltage, ground, and output. These are suggested since the need to build additional circuitry is not necessary.

The infrared region of the electromagnetic spectrum is quite large and runs on the low end at just above microwave radiation at 1mm or 1000 nm (300 GHz) to the high end at just below visible red light at about 700 nm (430 THz). As with other bands of the electromagnetic spectrum, IR is differentiated into subcategories. Infrared is closely associated with heat, with almost half of the sun's warmth reaching the earth coming in the form of infrared, with most of the rest in the visible wavelengths. Ultraviolet radiation is more energetic but only accounts for a tiny percentage of the sun's heat but can cause skin damage. Even with infrared at a much lower energy level than visible or UV light, care must be taken since there is an energy associated with it, and it can slightly penetrate human tissue. As a common-sense example: with IR used as a source for fiber optic communication, one should never look directly into the path of an end of a connection, or eye damage could result.

A video cameras' response to IR can be enhanced and used for night vision. There are also a great many sensors on the market for infrared detection, with the most common being the Passive Infrared (PIR) detector. It is sensitive to the movement of a heat signature and can activate alarm systems and various other devices.

To use infrared as a communications medium, one could vary the intensity of a transmitter in an analog manner to transfer analog signals, but the receiver will greatly be susceptible to interference by unwanted IR signatures. While there may be a purpose for this communication technique over a very short distance, it is far better to use a modulation scheme. To further eliminate interference, it is most common to use digital coding of a carrier signal. An example of digitally modulating an IR carrier is shown in the Oscilloscope screenshot of Figure 2.1. It is possible to vary the IR carrier in an analog manner but the circuitry for producing a square wave is far simpler.

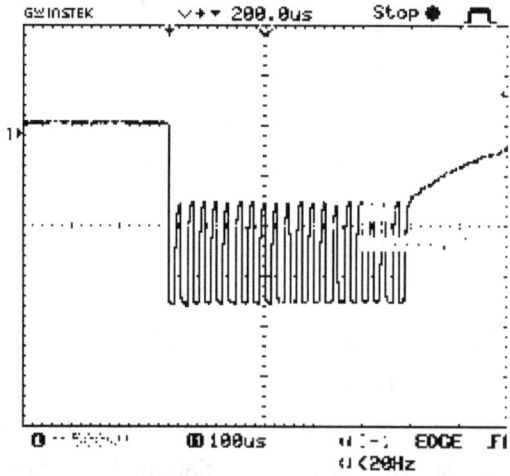

Figure 2.1. Digital modulated remote-control pulse

The Oscilloscope display shows one digital pulse of a pulse train being used to control a television. The 38 kHz carrier can be seen running for five blocks, with each block on the scope adjusted to equate to 100 us. The pulse we are viewing is 500 us in duration and is part of a pulse train that is not in view. The televisions' remote receiver would filter out the 38 kHz carrier and provide a clean 500 us digital coded pulse to the receive units' microcontroller. The presence and absence of pulses in a sequence determine a code in a Pulse Code Modulated system (PCM). Each coded pulse train can activate a function on the unit, such as changing channels, turning the power on/off, and so forth. IR remote control receivers are readily available at a low cost for maker projects. You can select a result to occur by differing the digitally modulated pulse width, or with a coding scheme for a pcm pulse train. We will experiment later in the chapter.

Section 2.2. IR advantages and limitations

Infrared communication devices are quite popular for close-range control of devices but are quickly being replaced by other technologies with more functionality and extended range. One example is Bluetooth, now Bluetooth low energy BLE. It outperforms IR since it can transmit through objects and has a typical range of up to 30 feet. If any more distance is needed, RF, or Wi-Fi modules can be utilized. We will investigate those technologies and protocols later in the text. The low cost and simplicity of IR still may make it the selected candidate for specific jobs. There are even workarounds when an IR signal must go through objects or over long distances. Some companies offer extenders when a device, such as a television, DVR, or cable box, must be controlled from an IR remote at a distance. An RF remote would immediately solve the problem, but RF extenders are available if an IR remote must be used with a unit. The extender has two sections, with the first section located near the IR remote transmitter. It converts the transmitted IR pulses to RF. The second part of the extender is located on the other end, and it reconverts the extended RF pulses back to an IR signal to reach the receiver.

A line-of-sight IR signal can travel very far if high-power IR LEDs are used to transmit the signal. They are readily available from manufacturers and can have current well up to 100 or so milliamps (ma) for maker projects. Transmit distance can also be enhanced by limiting the beamwidth for directionally. Some IR LEDs can be found with a beamwidth of only 10 degrees. For added range, the transmit LEDs can be wired in parallel so that an array can be utilized to transmit a more powerful IR signal.

Extremely high-power LEDs are available to aid in night vision for covert operations. LED arrays and lamps for vehicles can transmit infrared light at upwards of 60 watts, but these devices are for night vision equipment and not for communications. Lasers can be used for communications at very long distances but are limited due to their narrow beam and are best suited for distant fixed line-of-sight communications in the open air. This sort of application is used very infrequently. Lasers are, however, used in fiber optic communications.

Infrared signals can reflect off objects and is how clandestine cameras and night vision equipment can be augmented. Living objects produce heat with energy in the infrared range, which can be detected, but by flooding an area with infrared light which is not visible to the naked eye, the site can be further illuminated to show far more detail. Even at low power, the reflectivity property of IR can be useful, and many types of sensors are available. Examples are the automatic valve sensors used in kitchens and restrooms and some hands-free automatic door sensors. A later project will demonstrate the reflectivity of IR light.

Section 2.3. Lasers and fiber optics

We have all heard that white light is composed of all colors of the rainbow; it has a wide bandwidth range. Each color would have a small part of that bandwidth. LEDs used for communication have a typical bandwidth of 40 nanometers (nm), whereas laser diodes have a very narrow bandwidth of approximately 1 nm. Having a narrow bandwidth is an important consideration when working with fiber optic communication. When we speak of bandwidth, we are referencing the amount of frequencies required on the electromagnetic spectrum.

As was mentioned earlier, an LED or laser could be used for direct communication, but they have the limitation of being line-of-sight. That problem is overcome in a fiber optic cable because of Total Internal Reflection (TIR). If the illumination source is directed at a less than a critical angle into the cable, the cable can be bent, and virtually all of the light will exit on the other side. This is explained mathematically by Snell's Law: $n_1 \sin Q_1 = n_2 \sin Q_2$ which describes how light is refracted (bent) as it encounters optical material of different densities. The letter n in the formula is used for the refractive index of the optical material, and is given by the speed of light in free space, divided by the speed in the material. The letter Q represents the angle to the interface of the materials. Anyone who has poked a fishing pole into a lake has observed the effect. If a light source is pointed somewhat directly into the fiber cable, into what is sometimes called the acceptance cone, it can travel the length of the line. Even though fiber optic cable is made of extremely pure glass, some imperfections will lead to some attenuation of the light. On long-distance runs of fiber, repeater units are used to regenerate the light along the way. A narrower band light source can help avoid another problem when transmitting light through a cable called dispersion. It occurs due to the unequal time it takes for the slight variation of the waves bouncing from a wider band source to reach the output in multimode fiber. Dispersion distorts the signal. Graded index fiber resolves this situation by using a slightly denser center. Single mode fiber, however, is thinner and generally used on long distances. Fiber optic cable is surrounded by a material of different density, called cladding. The cladding is the second of the two densities shown in Snell's Law and gives us the internal reflection, but sometimes it can also exhibit dispersion as some wavelengths travel along the cladding. The plastic coating on the outside of the fiber optic cable is a protective jacket to decrease the chance of nicking the cladding, which could cause light leakage from the sides of the damaged fiber. The fiber optic cable shown in Figure 2.2 is from a cable TV company. Each tiny fiber optic cable is jacketed and is bundled inside of another jacket containing many lines. Ultimately, many bundles can be contained within a large protective jacket. The string running alongside the cables is made of Kevlar to provide mechanical strength since the line is hung from telephone poles.

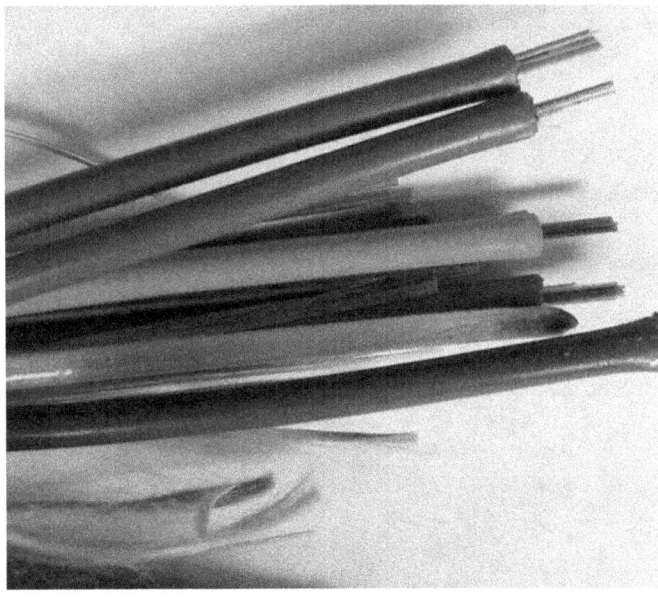

Figure 2.2. Cable TV Fiber optic cable

All diodes produce photons as electrons move within the device and go from a higher energy state to a lower one. LEDs and laser diodes are designed specifically to capitalize on this effect. Due to budget concerns, IR LEDs are a popular transmission source for shorter runs of fiber. Laser diodes have a higher failure rate and may require thermal stabilization. Lasers are preferable, however, because of their narrower bandwidth and faster operation. The devices operate by sending discrete light pulses, and LEDs may take as long as 5 nanoseconds (ns) to switch states. In contrast, a laser is under 1 ns and can enable far faster communication. Pin or avalanche diodes are used as detectors on the receiving end.

Section 2.4. IR light reflection project

We will use an IR LED to transmit a pulse which, when reflected by an object, will activate a visible LED. The first section of this project will utilize a NE555 timer set up in astable mode to pulse at the approximate 38 kHz frequency necessary to activate the receiver when a reflection is detected. The receiver is a TSOP4838 containing a pin diode and preamplifier. It is oriented with the rounded dome section facing towards you as described in the pin-out of Figure 2.3.

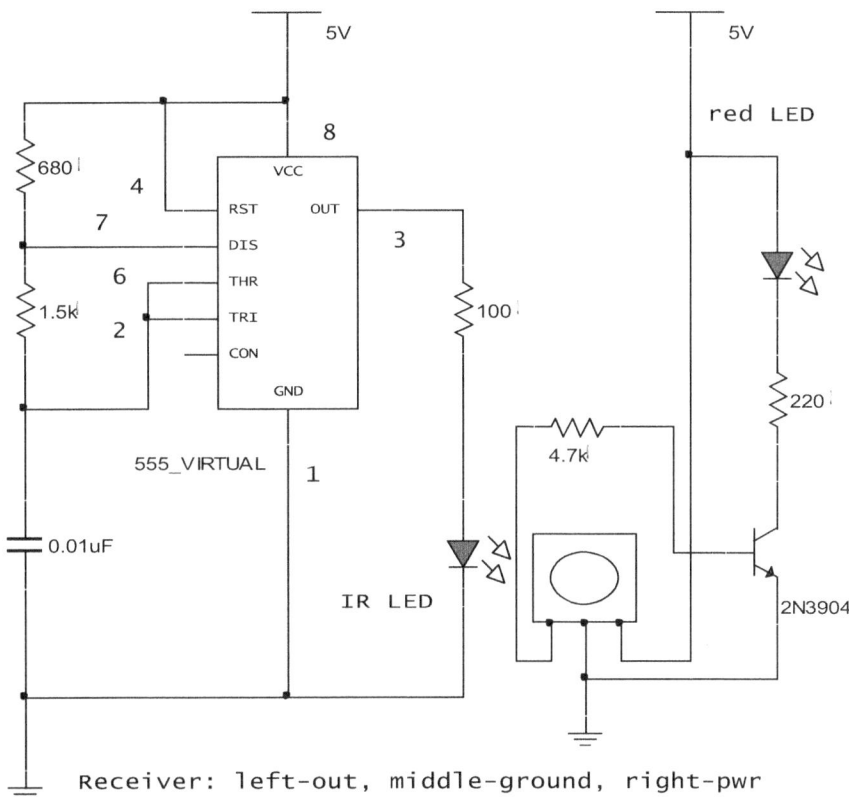

Figure 2.3. IR reflection schematic

The 555 timers were first commercialized in the early 1970s and have been one of the most popular ICs for maker projects. The schematic shows a 680 Ohm and a 1.5 k Ohm resistor along with a 0.01 uF capacitor used for timing. For better accuracy, a potentiometer could have also been added since the 555 frequency is approximate due to the values of component tolerances. In testing the circuit shown in Figure 2.3, I found a pulse frequency of 37 kHz, which is within the bandwidth of the IR receiver, and produced a change in the visible LED. I was able to cause an IR reflection with the palm of my hand from about 5 inches away from the bottom side of the breadboard. You can also bounce the signal from a white sheet of paper to extend the reflection. The range would be extended if the frequency were adjusted closer to the center frequency of 38 kHz. The TSOP4838 can also be upgraded to a receiver module, such as the DRF0094. Additionally, the IC output could bias a driver transistor to operate a higher power IR LED using a smaller current limiting resistor. A representative circuit is shown on a breadboard in Figure 2.4.

Figure 2.4. IR reflector circuit

The orientation of the NE555 pin one is on the lower left-hand side. The IR LED and the LED dome section of the TSOP4838 are facing in the same direction. There should be a small sheet of thick paper or some other opaque object between the transmitter and receiver during operation to prevent any IR from leaking from the side and reaching the receiver directly. The 2N3904 NPN transistor is located to the right of the IR receiver. The IR receivers' left-hand side pin is the output which is normally at a high level with no signal, biasing the driver transistor to conduct, thus lighting the visible LED in the collector circuit. The LED current is limited by the 220 Ohm resistor in the collector circuit. When an IR signal is detected, the receivers' output goes low and extinguishes the visible LED. The transistor circuit is used not to overdrive the receiver.

An example of the operation as we have described could be used to stop a mobile robot or other vehicle when an obstacle is detected in its path. This is demonstrated when a reflection is received, and the LED turns off. This project can also be adapted to have the LED normally extinguished and to only light when a reflection occurs by exchanging the 2N3904 NPN transistor with a 2N3906 PNP transistor. The PNP device would need to be reversed from the NPN that is drawn in the schematic. In the pictorial view of Figure 2.4, the driver transistor shown to the right of the receiver would simply be reversed so that its rounded side would be facing toward the front edge of the breadboard where the IR reflection is taking place. Using a PNP transistor in this way will light the LED when an IR reflection occurs and is a bit more intuitive.

Section 2.5. IR direct line-of-sight project

There are several options for this project. We could continue using the NE555 IR transmitter circuit from Figure 2.4, and relocate the receiver section to a different breadboard, or replace the 555 and use an Arduino to generate the 38 kHz transmitter pulses. If you had poor range with the last project, you might redo the project using an Arduino to generate the pulses before continuing with this section. The Arduino, however, is only capable of generating a high-level output of 20 ma. We are using a driver transistor to provide current to the Transmit IR LED as shown in Figure 2.5, but the receiver circuit will remain as shown in the previous schematic Figure 2.3. You can lower the value of the 100 Ohm current limiting resistor in transmitter circuit Figure 2.5 to extend the range if you are using a high power IR LED.

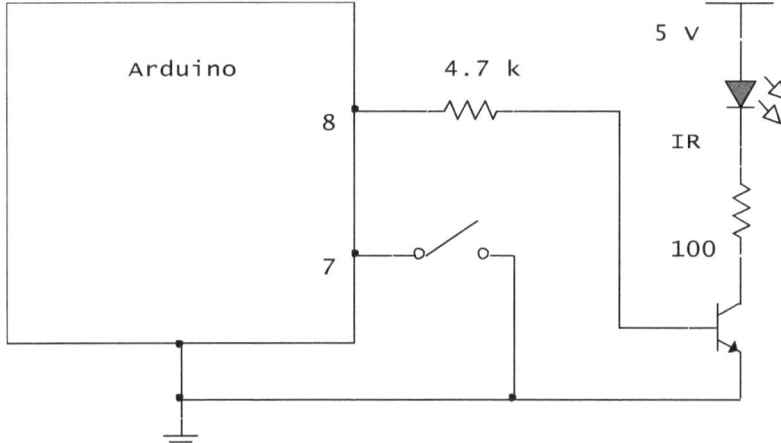

Figure 2.5. Arduino IR transmitter

The Arduino code is concise since we use the tone function to generate the driver pulses to the transistor. The code is shown in Code Listing 2.1.

```
const int IRout = 8;
void setup(){
  pinMode (IRout, OUTPUT);
}
```

```
void loop(){
    tone (IRout, 38000); // IR LED pulsing
    delay (5000); //5 second delay before looping
}
```

Code Listing 2.1. 38 kHz pulse generation

The five-second delay is unnecessary but eliminates some of the loop time delays. We show a momentary switch connected to pin 7 in Figure 2.5 if you wish to add an on/off function to the project code. We specified the 38 kHz frequency as a direct number, however, if you modify the code and use a variable to represent the frequency, it must be declared as an unsigned integer to accommodate the 38 kHz tone.

Section 2.6. IR control link project

A control link requires a separate transmitter and receiver. You can think of it as when sitting on a chair controlling a TV or some other device. When you press a button on the remote control, a signal goes to an IR receiver in the appliance where the pulse code is recognized and acted upon. While it would be possible to use a single Arduino controller and rely on close range reflected pulses in this project, it would be more realistic to work in a collaborative group using two units separated by a distance. You can start at a short distance and test the range. The drawing for the transmitter is shown in Figure 2.6.

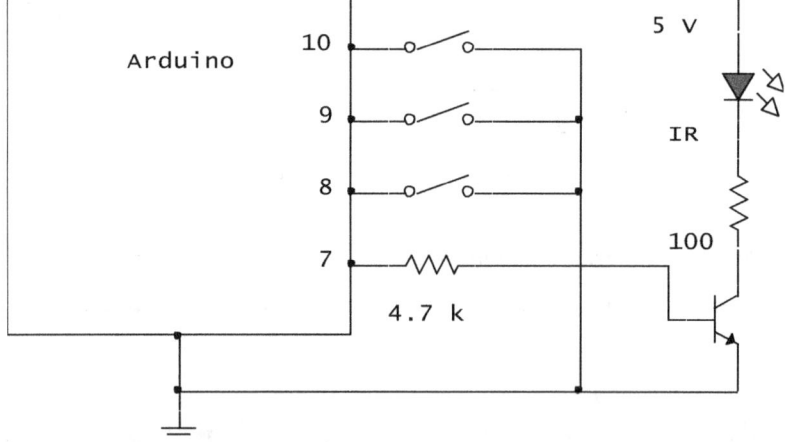

Figure 2.6. IR control link transmitter

The IR LED current is operated by a 2N3904 NPN transistor driven from pin 7 of the Arduino. A high output from pin 2 turns on the LED. The range can be increased by lowering the value of the current limiting resistor to less than 100 ohms, and or raising the voltage in the LED section of the circuit, if you are using a high-power IR LED. You make a selection by momentarily tapping a wire from either of pins 8, 9, or 10 to ground. A quick tap to the USB connector case works fine. Each pin is programmed to send ten sets of a repetitive pulse duration, each with different timing to specify a tone the receiver will sound. Part of the pulse train transmitted output for having the receiver sound a 500 Hz tone by selecting transmitter pin 8 is shown in Figure 2.7.

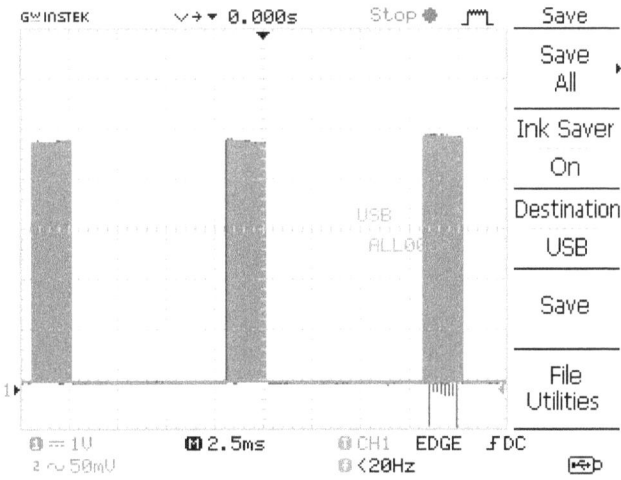

Figure 2.7. 10 ms transmit pulse cycle

The dark pulses in the oscilloscope picture consist of the 38 kHz carrier sent for 2 ms, followed by no transmission for 8 ms, giving a pulse repetition time of 10 ms and a pulse repetition frequency (PRF) of 100. The cycle repeats ten times, but we only have just over two cycles displayed in the figure. Code Listing 2.2 is the code for the transmitting Arduino controller.

```
//The code for the transmitting Arduino follows:
const int sound1 = 8;
const int sound2 = 9;
const int sound3 = 10;
const int IRout = 7;
int lowTone;
int medTone;
int highTone;
```

```
int sound;
int timer;
void setup() {
  pinMode (sound1, INPUT_PULLUP);
  pinMode (sound2, INPUT_PULLUP);
  pinMode (sound3, INPUT_PULLUP);
  pinMode (IRout, OUTPUT); // to IR LED section
}
void loop() {
  lowTone = digitalRead (sound1);
  medTone = digitalRead (sound2);
  highTone = digitalRead (sound3);
  if (lowTone == LOW) {
    sound = 1;
  }
  if (medTone == LOW) {
    sound = 2;
  }
  if (highTone == LOW) {
    sound = 3;
  }
  switch (sound) {
    case 1:
      while (timer < 10) {
        tone (IRout, 38000);//outputs a short duration 38,000 tone
        delay (2);
        noTone (IRout);
        delay (8); //total period 10 ms
        timer = timer + 1;
      }//sends 10 times
      break;
    case 2:
      while (timer < 10) {
        tone (IRout, 38000);
        delay (2);
        noTone (IRout);
        delay (18); //period 20 ms
        timer = timer + 1;
      }
```

```
      break;
    case 3:
      while (timer < 10) {
        tone (IRout, 38000);
        delay (2);
        noTone (IRout);
        delay (28); //period 30 ms
        timer = timer + 1;
      }
      break;
  }
  sound = 0;
  timer = 0;
}
```

Code Listing 2.2. IR Link, transmitter code

After the I/O pins are named, and the variables are defined, the setup section runs once and denotes I/O. The main loop examines the inputs and responds to a low level when a pin is grounded. The *switch case* conditional code is similar to using *if/then* statements but streamlines the process. Each case will transmit a short pulse with a long time delay. Ten cycles make up the pulse train. Only two are needed, and the rest help with the reliability of the transmission. The TSOP4838 or equivalent receiver will output a low level while receiving the 38 kHz carrier during pulse transmission and will otherwise be at a high output level. The distances between pulses are interpreted as a command to generate a specific tone as described in Code Listing 2.3.

```
//The code for the receiver Arduino board follows:
const int IRin = 7; // IR receiver module connect
const int spk = 8;
int pulseTone;
int pulsePeriod;
unsigned long currentTime;
unsigned long oldTime;
void setup() {
  pinMode (IRin, INPUT_PULLUP);
  pinMode (spk, OUTPUT);
}
void loop() {
```

```
  pulseTone = digitalRead (IRin);
  if (pulseTone == LOW) {
    currentTime = millis();
    pulsePeriod = currentTime - oldTime; //picks up the second burst
    oldTime = currentTime;
    if (pulsePeriod > 6 && pulsePeriod < 14) {
      tone (spk, 500);
      delay (5000);
      noTone (spk); //speaker shuts off
      pulsePeriod = 0; //reset
    }
    if (pulsePeriod > 16 && pulsePeriod < 24) {
      tone (spk, 1000);
      delay (5000);
      noTone (spk); //speaker shuts off
      pulsePeriod = 0;
    }
    if (pulsePeriod > 26 && pulsePeriod < 34) {
      tone (spk, 5000);
      delay (5000);
      noTone (spk); //speaker shuts off
      pulsePeriod = 0;
    }
  }//end of IRin reception
}
```

Code Listing 2.3. IR Link, receiver code

The Arduino contains a millisecond timer that begins counting when power is applied. At the beginning of the loop section, after the first pulse of the train has run, the code determines the width of each cycle, and if it fits one of the three pulse periods, one of three tones will be heard from the speaker in the circuit shown in Figure 2.8.

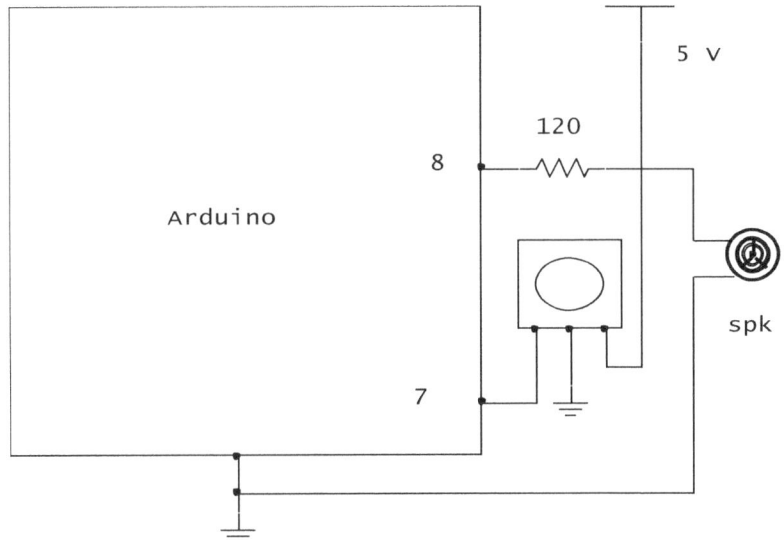

Figure 2.8. IR control link receiver

The audio tones we are using are the same as were used in the project of section 1.5. We again utilize the tone function to output the tone through the current limiting resistor to the speaker. The noTone command shuts off the tone after a five-second delay.

The amount of transmitted data is limited by the type of modulation we are using in this example. It is a version of pulse position modulation; however, for simplicity, our pulse cycles are quite large, which limits the number of commands which can be sent in a given amount of time. Most modern digital transmission links use a version of pulse code modulation (PCM). This system usually has a group of synchronizing pulses and sends digital ones and zeros in a packet to form a code that the receiver will interpret. Our next chapter examines more of these modulation techniques.

Chapter Two Summary

IR remote controls for home devices use a standard wavelength of 940 nanometers flashing at a square wave frequency of 38 kHz to produce carrier pulses. The carrier pulses can be modulated in many ways to send information through a link. Reliability from the elimination of unwanted infrared sources is the reason that a carrier is used. There is no reason why analog information cannot be sent directly and may be a good project for an adventurous maker, but digital signals are far easier to generate, and the receivers are readily available. IR receivers like the TSOP4838 and modules, such as the DRF0094 contain amplification circuits; however, standalone IR receiver LEDs are also available for those wishing to build amplifiers. Direct infrared is like visible light in that it tends to travel in straight lines but can refract and reflect. Fiber optic cables are used for long distances but must have repeaters. They can be spread out to significant distances are far as 50 miles on high-power long-distance systems. Fiber optics systems can operate at extremely high speed and high capacity when multimode cable is used for short distances, and single mode for the longer runs. Multimode fiber signals can reflect along different pathways and distort the output pulse, which is called dispersion.

Chapter Two Questions

1. Why is it important to use carrier pulses on an infrared communications system?
2. What is the standard frequency of IR carrier pulses?
3. Which of the following are below the infrared range of the electromagnetic spectrum: visible light, microwave energy, ultraviolet light?
4. Give an example of where the reflectivity of an IR pulse would have a commercial application?
5. What mode of operation would a 555 timer be used to generate periodic pulses?
6. What is the name of the mathematical law that describes how light is refracted (bent) as it encounters optical material of different densities?
7. Why is it important to use a standard wavelength for infrared devices?
8. Explain what the cladding is for in a fiber optic cable.
9. Why is Kevlar sometimes used in fiber optics?
10. When using the tone function in generating sounds with the Arduino, what command shuts off the sound?

Chapter 3

Bluetooth, Wi-fi, and ISM devices

Section 3.1. Bluetooth and BLE

Bluetooth generally only requires one transceiver module since a control app is typically loaded on a smartphone. There are dual master/slave modules available if you want to produce a stand-alone project. The original Bluetooth range was very short, at only about 10 feet. Later standards range nearly 30 feet under ideal conditions, and over 100 feet is now possible with the latest popular standard, Bluetooth 4.0. (Bluetooth 5.0 promises a range of upwards of 1000 feet and uses Wi-Fi frequencies.) The ranges are variable since the transmissions are at the high end of the UHF band and communications tend to be line-of-sight at such high frequencies and easily blocked by objects. Objects in the path can block or attenuate high-frequency transmissions, whereas lower frequency signals may not be affected and can bend around the obstructions.

We covered Industrial, Scientific, and Medical (ISM) allocations in Chapter One. There are many frequencies set aside for equipment that generate RF signals due to the nature of their operations and can cause interference to communications. Licensed radio services are not usually assigned to these frequencies on a primary basis but can share specific ISM frequencies. Unlicensed low-power ISM communications can take place but are subject to interference from other devices. This drawback was not an issue in the past since Bluetooth was initially designed for close-range use in earbuds and mobile phone links to nearby computers. The typical maximum RF power output for Bluetooth is 25 milliWatts (mW) but can go higher, with up to 100 mW in the latest specification. Bluetooth can use a frequency hopping technique to spread its transmissions across the band to minimize the amount of interference. This technique is called spread spectrum communication.

Frequency hopping can be both passive or active. Active frequency hopping uses a receiver to sweep across the frequency range in military applications and other communications subject to being jammed. Then the transceivers are coordinated to utilize the quietest frequency within the range. If the jamming station comes on frequency, then the process is repeated. This method can also happen at specific periods to aid in coordination. The frequency hops are preplanned and may be pseudo-random in a passive spread spectrum design regardless of the channel's noise. A passive system is much simpler to design. Timing both receivers and transmitters to be on the same channel simultaneously is imperative for a link. Original Bluetooth sends data packets over 79 channels, with each channel having a bandwidth of 1 MHz. Under the popular 4.0 standards, Bluetooth Low Energy (BLE) uses 40 channels spaced 2 MHz apart, and low energy usage increases battery life.

All RF transmissions are analog sine waves but can be modulated digitally for greater information exchange. There are three main ways to modulate analog RF carriers. They are by either modifying the RF amplitude, frequency, or phase. Sometimes these methods are combined. The first digital modulation that was popular in Bluetooth devices was Gaussian Frequency Shift Keying (GFSK). Regular Frequency Shift Keying (FSK) sends binary ones and zeros by slightly moving the carrier higher or lower in frequency to represent the bit status. A one in a digital transmission is referred to as a mark and a zero as a space. Another popular digital modulation technique is Phase Shift Keying (PSK). Rather than changing frequency to represent the bit level, only the signals' phase is shifted higher or lower. GFSK usually employs a filter with a Gaussian response to remove unwanted sideband emissions generated from modulating the carrier. Work continues on changes to the modulation schemes to achieve higher bit rates and a more reliable range.

There are many uses for Bluetooth devices in the Internet of Things devices. Bluetooth uses a Master/slave protocol. Pairing for special applications has been simplified, but in a project that we will later build, we will use a phone app to scan for nearby Bluetooth devices and enter a code to connect. We will be using the popular HC-05 module shown in Figure 3.1. It is shown on the left side next to the 4.0 version HC-10, on the right. For pairing a module to a phone, the default code with the HC-05 wiil be either 0000 or 1234.

Figure 3,1. Bluetooth Modules HC-05 and HC-10

The rectangular zig-zag area at the top of the modules is the transmit and receive antenna. Pin spacing at the bottom is breadboard compatible, but the module requires input data levels at 3-volts, which we will compensate for in our project by using a voltage divider. The HC-10, shown on the right, looks very similar but has a far greater range of nearly 100 meters and lower power consumption. It is built to Bluetooth 4.0 standards and operates in the 2.5 GHz ISM band. Its data rate is 24 Mbs as opposed to the HC-05's 3 Mbs and its 30-foot range. The programming is slightly more challenging with the HC-10, but it may be appropriate if you need the enhanced operation for your next maker project.

Section 3.2. Wi-Fi

Wi-Fi was initially used in short-range LAN links to provide wireless Ethernet in networks but has taken on added dimensions now with the Internet of Things (IoT) and other uses. The two most common RF bands for today's Wi-Fi are 2.4 GHz and 5 GHz, but there are many other bands in use, as well. Channel spacing is 5 MHz, but some devices can simultaneously use multiple channels for greater throughput. Just as with Bluetooth, the line-of-sight characteristics of these high frequencies tend to be blocked or attenuated by objects in the transmission path, and ranges can vary significantly. And, like Bluetooth, because Wi-Fi stations produce unlicensed transmissions in ISM bands, they are subjected to interference. Devices are given a Service Set Identifier (SSID) and can use MAC addresses assigned by equipment manufacturers. Each country sets its permissible maximum power output. In the United States, rather than give Wi-Fi a power measured at the transmitter output. Effective Radiated Power (ERP) is used and takes antenna gain into the equation. With standard gain antennas, the typical Wi-Fi range is 100 meters. With highly directional high-gain antennas, the range could be extended to many kilometers in a line-of-sight deployment. On low power installations without special antennas, repeaters can be used to extend coverage. The modulation in different Wi-Fi standards are variations of Phase Shift Keying (PSK), usually coupled with Amplitude Modulation (AM).

There are many Wi-Fi low-cost modules available for maker projects. One such module is the NRF2401. They require 3-volt power supplies, consume very low power, and operate on 125 channels in the 2.4 GHz ISM band. An NRF2401 and an add-on power regulator module are pictured in Figure 3.2.

Figure 3.2. NRF2401 and power module

The NRF2401 Wi-Fi transceiver is shown on the right, with its antenna located on the right edge of the board. It is a highly integrated device. The elliptical component on the bottom edge is the crystal resonator used for frequency generation. The only other elements are the IC and a few surface mount resistors and capacitors. The board runs on 3-volts and has a nonconforming pin orientation for breadboard use. The other board shown on the left contains a 3-volt regulator and a standard set of pins for breadboard use. The additional module is helpful but not necessary for NRF2401 operation. They are not set up for Internet Protocol but using two of the transceivers can provide a communication link for Maker projects. The coding is somewhat complicated and requires downloading libraries, but many projects, codes, and videos are available online for these devices.

Section 3.3. ISM devices

Earlier in the text, we introduced you to a low-cost 433 MHz receiver and transmitter unit, as shown in Figure 3.3.

Figure 3.3. 433 MHz transmitter and receiver

Both the Bluetooth and Wi-Fi devices we discussed in the chapter have been transceivers. A transceiver contains both transmitter and receiver in one package. The 433 MHz devices in Figure 3.3 are separate, with the receiver shown on the right and the transmitter pictured on the left. The receiver is a superheterodyne variety somewhat like a receiver found in an AM radio; however, this receiver only outputs digital logic levels. The transmitter uses AM with a communications modulation called Amplitude Shift Keying (ASK). The carrier is switched on and off to reflect the data levels. When the data to be sent is low, the carrier is off, and a high logic level transmits the carrier for the duration of the logic pulse.

The transmitter can be operated up to 12 volts for VCC with data levels at zero volts representing a low and a maximum of five volts for a high. The receiver VCC can go no higher than five volts. Both modules must have a 17.3-centimeter wire antenna attached to a solder connection on each of the boards. This length is one-quarter of a wavelength at the 433 MHz operating frequency. It can be coiled at the base to reduce the overall length. Figure 3-3 shows that the input data pin for the transmitter is on the left (it is stenciled up-side-down), VCC is in the center, and the ground pin is on the right. VCC is on the left for the receiver, both middle pins output data, and the ground pin is on the right. This pinout is typical for the modules, but several versions exist, and you should look for the pinout stenciled on each board or listed in the datasheet.

The 433 MHz, ISM band, has quite a bit of noise from other devices like home security units, temperature sensors, and a wide range of different devices, so the range is limited and variable. Under good conditions, it is possible to transmit up to 50 meters in an open area on average. We will use this transmitter and receiver in one of our later projects.

Section 3.4. Bluetooth remote temperature sensor project

We will be using an unusual electronic component, a specialized variable resistor called a thermistor, whose resistance varies with temperature. The part number is NTC MF52-103, a 10 k Negative Temperature Coefficient (NTP) thermistor; as temperature goes up, resistance goes down. (A negative coefficient produces an inverse result.) We are placing it in a voltage divider with a fixed value 10 k resistor, as shown in the diagram of Figure 3.4.

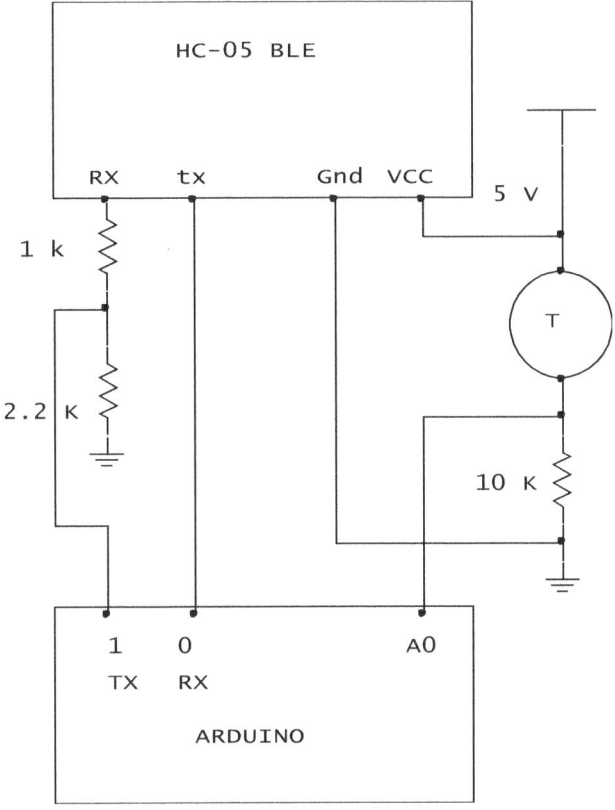

Figure 3.4. Bluetooth remote temperature sensor

Regular conductors, such as wires, have a positive temperature coefficient (PTC). As the current in a wire increases, it produces more heat from an increase in the random vibrations of electrons, limiting controlled current flow. On power distribution systems, this produces a loss of energy called an I^2R power loss. Thermistors, however, are made of semiconductor material. The NTC MF52-103, 10 k thermistor is physically very small and pictured in Figure 3.5 against a standard notebook page background.

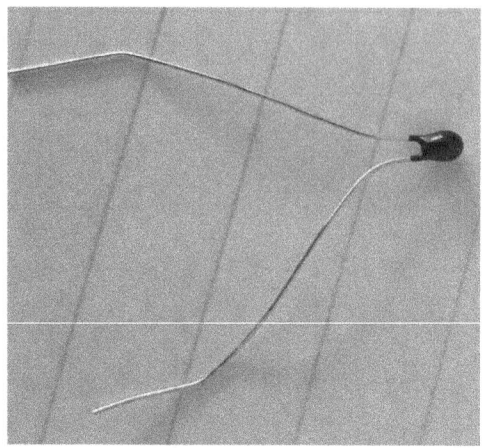

Figure 3.5. 10 k Thermistor

Typical semiconductors such as diodes and transistors usually have a negative temperature coefficient like the thermistor. But a NTC can have detrimental results. This effect can cause thermal runaway in high-powered audio amplifiers and computer IC chips. As the temperature goes up, resistances go down, and cooling becomes extremely important to keep components from overheating and is a reason why you can hear more activity from cooling fans as computer processors run more vigorously. In the manufacture of thermistors, most have a NTC, but some have a PTC for use in specialized applications like self-regulating heating devices. NTC thermistors are mainly used for temperature measurements, like the one we are using in this project. Their resistance value is specified at an ambient temperature of 25 C. They are nonpolarized. The breadboard picture is shown in Figure 3.6.

Figure 3.6. Remote temperature sensor breadboard circuit

The Arduino has a beneficial feature that can be useful for troubleshooting called Serial.print. It allows you to print a message to the serial monitor screen to show the progress of a program as it is running. It also can be used to respond to a user interactively. We will be using the Serial.print function to transmit written information to either a cellphone or to the computer screen. This first project has minimal user input and only produces an output of the current temperature in the phone app, but we will examine a more interactive activity using Bluetooth in our next project. The code for this project is shown in Code Listing 3.1. (The Bluetooth TX and RX pins must **not** be connected during the upload or there will be a serial port conflict.) After the program is entered in the Arduino and ready to connect to the smartphone, the default code must be entered into the app, which is usually 0000 or 1234.

```
float degreeIn;
float degree;
float degreeTotal;
float degreeAve;
int degreeC;
int degreeF;
int check;
void setup() {
  Serial.begin(9600);
}
void loop() {
  while (check < 8) {
    check++;
    degreeIn = analogRead(A0);
    degree = log(((10240000 / degreeIn) - 10000));
    degree = 1 / (0.001129148 + (0.000234125 * degree) + (0.0000000876741 * degree * degree * degree));
    degreeTotal = degreeTotal + degree;
    delay(500);
  }
  if (check >= 8) { //resets loop at 9
    check = 0;
  }
  degreeAve = degreeTotal / 8; //8 readings
  degreeTotal = 0; //reset
  degreeC = degreeAve - 273.15;        // Kelvin to Celsius
```

```
    degreeF = (degreeC * 9.0) / 5.0 + 32.0; // Celsius to Fahrenheit
    Serial.print (degreeC);
    Serial.println (" C");
    Serial.print (degreeF);
    Serial.println (" F");
    Serial.println ("");
    delay(1000);
}
```

Code Listing 3.1. Remote temperature sensor code

The float type variable is used in the Arduino code to represent fractional numbers. The serial port is enabled at a 9600 baud rate, which is slow but sufficient for text and is a standard speed. The analog voltage at the junction between the 10 k thermistor and 10 k resistor is sampled and sent to an Analog to Digital Converter (ADC) in the controller. For a reliable and steady reading, the voltage at the thermistor is sampled eight times over a four second interval and the average is found. The math in the code first finds the temperature on the Kelvin scale, then converts it to both Celsius and Fahrenheit before sending it to a Bluetooth app on a smartphone.

Notice the formatting in the screen print section, spaces between characters appear on the reader and Serial.println generates a new line. (The letters after "print" are lowercase L and N.) A screenshot in Figure 3.7 is of the program in operation as it appears on a Samsung phone using the Android 8.1.0 operating system.

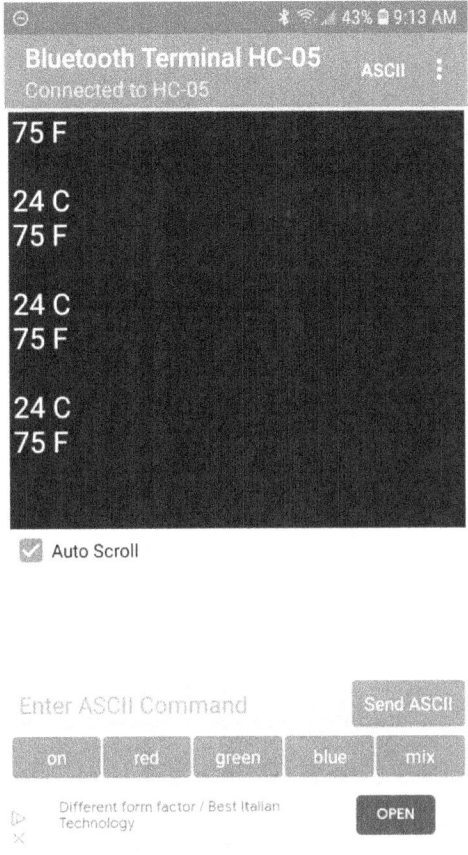

Figure 3.7. Remote temperature sensor program operation

There are many Bluetooth apps available, and many are free. We are using "Bluetooth Terminal HC-05" by mightyIT, which can be readily downloaded. The free app shows periodic ads. The professional version is inexpensive and not supported by advertising. Most Bluetooth apps require location permission and must be paired. In our case, the code for the device is 1234. A 9-volt battery snap is shown in Figure 3.6, plugged into the barrel connector for power to the Arduino. The 5-volt out and ground pin supplies power to the breadboard. These connections allow remote operation without a computer. For troubleshooting purposes, you can remove the battery and Bluetooth module, and then connect the Arduino to the computer via USB. (Don't forget to select the highest port number.) The serial monitor should display the temperature readouts. The serial monitor is accessed under the Tools menu and can also be accessed from the magnifying glass icon in the top right of the IDE, under the close program "X", as shown in Figure 3.8.

Figure 3.8. Serial Monitor icon

The accuracy may be slightly improved by using a 15 k variable resistor and calibrating the thermistor voltage divider. The code may also be modified to keep the temperature display at a steady reading throughout a number of display cycles. Another great Maker project might use a temperature and humidity sensor, as shown in Figure 3.9.

Figure 3.9. DHT11 Temperature and humidity sensor module

The DHT11 sensor is very accurate and readily available. In the figure, it is in module form, but the actual sensor is the box located on the right-hand side of the figure. If it is used without a module, some additional components are required on the breadboard. The following section will demonstrate even more interactivity with Bluetooth.

Section 3.5. Bluetooth controlled lighting project

In looking closely at the screenshot of program operation in Figure 3.7 of the last section, several buttons are visible at the bottom of the picture and are listed as On, Red, Blue, etc. The buttons are for user interactivity, and we will use them to create an outdoor lighting effect in this section. The goal of the project is to provide an exciting area of outdoor landscape lighting but can be used in many different applications. An outdoor four-color lamp assembly can be purchased and modified at a home renovation store like Home Depot or produced in a home workshop. A picture of the type of outdoor lamp we are using is shown in Figure 3.10.

Figure 3.10. Multicolor outdoor lamp

The "On" button in the phone app in Figure 3.7 was changed to "White" along with the corresponding ASCII code. During operation, a solid white light appears when power is first applied. The color buttons produce the appropriate color output, and the "mix" button will cause a continuous alternation of the colors. Pressing any button will cancel the previous operation and begin the new one. I used an Arduino Nano for this project because the Nano fits directly into a breadboard, but the Arduino Uno will work just as well. The breadboard circuit is encased in a waterproof container located vertically connected to a rafter under an outdoor deck, as shown in Figure 3.11. Since professional cases are costly, a Tupperware container was used to protect the controller from the elements. It is displayed with the cover removed.

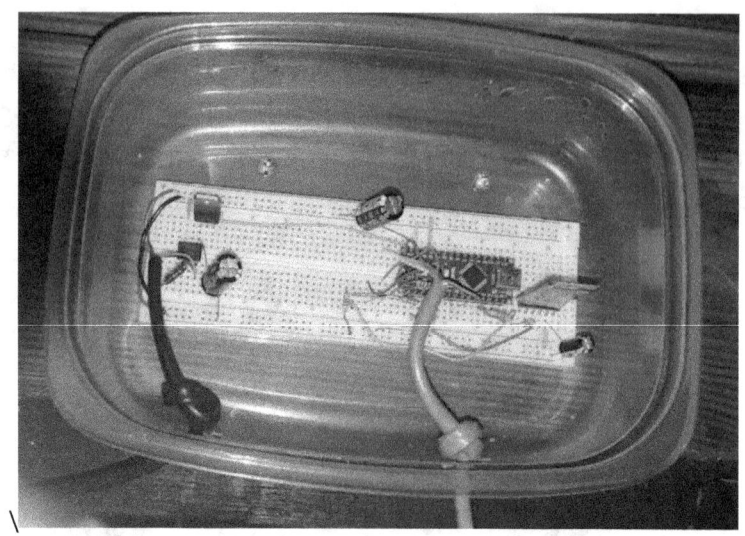

Figure 3.11. Landscape lighting project breadboard

    A two-section breadboard was used. Some two-section versions will split the power busses which run along the edges of the board. It is always a good idea to test continuity between the far sides of the edge power busses on this type of breadboard. If the busses are split in the middle, small jumper wires may be installed to connect the two sides, if necessary.

    Since most low-power landscape lighting runs on 12 VAC, a small power supply can be seen on the left of Figure 3.11, consisting of a bridge rectifier and a 9-volt regulator with a filter capacitor connected to the "power in" pin of the Arduino Nano. The wiring shown next in Figure 3.12 is used with an Arduino Uno and single section breadboard with four standard LEDs.

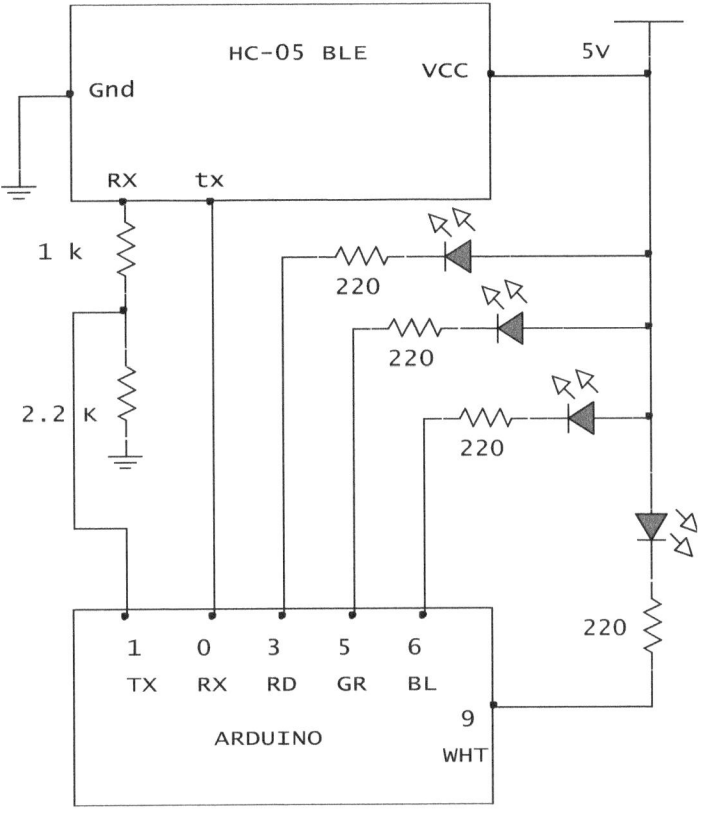

Figure 3.12 Landscape lighting circuit

The Bluetooth module and the phone app are in two-way communication. By coding the app buttons, the phone will send the associated ASCII characters which are recognized in the code, to perform the actions. Decimal ASCII code numbers represent all of the functions of a computer keyboard. Each is converted into an 8-bit number for processing. The app buttons are very straightforward to use, as shown for the red button in Figure 3.13. Holding them steady will produce an expanded view available for programming, as we see in the picture. We are using the free version of the app, so you may also notice periodic advertisements.

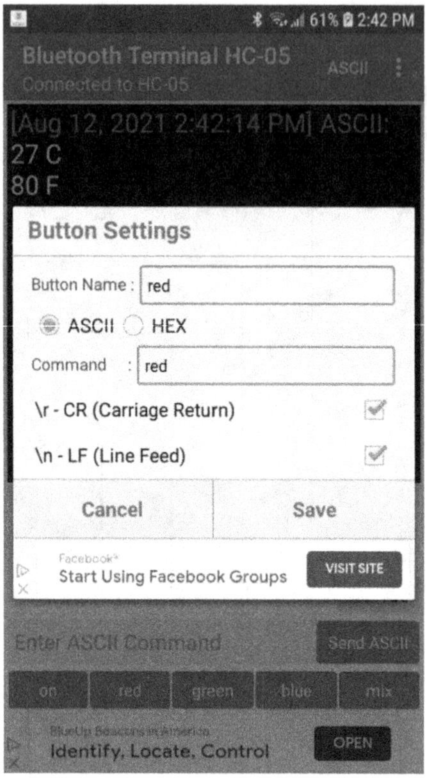

Figure 3.13. Programming the app buttons

We type the label for the button that the user will see and then select the ASCII selection. The codes which represent the letters will be sent to the Arduino when the button is selected. The Arduino code for operation is shown in Code Listing 3.2. (The Bluetooth TX and RX pins must **not** be connected during the code upload.)

```
char c;
byte d;
byte x;
byte action;
byte arrayCall[8]; //input array
byte arrayWhite[8];
byte arrayRed[8];
byte arrayGreen[8];
byte arrayBlue[8];
byte arrayMix[8];
const int red = 3;
const int green = 5;
const int blue = 6;
```

```
const int white = 9;

void setup() {
  Serial.begin (9600);
  pinMode (white, OUTPUT);
  pinMode (red, OUTPUT);
  pinMode (green, OUTPUT);
  pinMode (blue, OUTPUT);

  digitalWrite (white, LOW); //high deactivates LEDs on the lamp unit
  digitalWrite (red, HIGH);
  digitalWrite (green, HIGH);
  digitalWrite (blue, HIGH);

  arrayWhite [0] = 119; //w //Button White
  arrayWhite [1] = 104; //h
  arrayWhite [2] = 105; //i
  arrayWhite [3] = 116; //t
  arrayWhite [4] = 101; //e

  arrayRed [0] = 114;//r //Button Red
  arrayRed [1] = 101;//e
  arrayRed [2] = 100;//d

  arrayGreen [0] = 103; //g //Button Green
  arrayGreen [1] = 114; //r
  arrayGreen [2] = 101; //e
  arrayGreen [3] = 101; //e
  arrayGreen [4] = 110; //n

  arrayBlue [0] = 98;//b //Button Blue
  arrayBlue [1] = 108;//l
  arrayBlue [2] = 117;//u
  arrayBlue [3] = 101;//e

  arrayMix [0] = 109;//m //Button Mix, toggles fast, med, slow, and very slow
  arrayMix [1] = 105;//i
  arrayMix [2] = 120;//x
}//end of setup
```

```
void loop() {
  while (Serial.available()) {
    delay(70);
    byte d = Serial.read(); // get ASCII Byte
    char c = d; //convert to char
    Serial.print(c);
    if (d > 13) {//skips new line and carrage return etc.
      arrayCall[x] = d; //loads call memory
      x++;
    }
  }// end of serial available
  for ( x = 0; x < 8; x++) {
    if (arrayCall[x] == arrayWhite [x] && arrayCall[x] > 13 ) { //action 5, white call
      action = 5;
    }
    if (arrayCall[x] == arrayRed[x] && arrayCall[x] > 13 ) { //action 4, red call
      action = 4;
    }
    if (arrayCall[x] == arrayGreen [x] && arrayCall[x] > 13  ) { //action 3, green call
      action = 3;
    }
    if (arrayCall[x] == arrayBlue [x] && arrayCall[x] > 13 ) { //action 2, blue call
      action = 2;
    }
    if (arrayCall[x] == arrayMix [x] && arrayCall[x] > 13 ) { //action 1, mix call
      action = 1;
    }
  } // end of for

  switch (action) {
    case 5: //white on
      digitalWrite (white, LOW );
      digitalWrite (red, HIGH);
      digitalWrite (green, HIGH);
      digitalWrite (blue, HIGH);
      clearArray();
      break;
```

```
case 4: //red
  digitalWrite (white, HIGH);
  digitalWrite (red, LOW);
  digitalWrite (green, HIGH);
  digitalWrite (blue, HIGH);
  clearArray();
  break;

case 3: //green
  digitalWrite (white, HIGH);
  digitalWrite (red, HIGH);
  digitalWrite (green, LOW);
  digitalWrite (blue, HIGH);
  clearArray();
  break;

case 2: //blue
  digitalWrite (white, HIGH);
  digitalWrite (red, HIGH);
  digitalWrite (green, HIGH);
  digitalWrite (blue, LOW);
  clearArray();
  break;

case 1: //mix
  digitalWrite (white, HIGH);
  digitalWrite (red, HIGH);
  digitalWrite (green, HIGH);
  digitalWrite (blue, HIGH);

  for (x = 255; x > 0; x--) {
    analogWrite(red, x);
    delay (10);
  }
  for (x = 0; x < 255; x++) {
    analogWrite(red, x);
    delay (10);
  }                    //red up/down
```

```
    for (x = 255; x > 0; x--) {
      analogWrite(green, x);
      delay (10);
    }
    for (x = 0; x < 255; x++) {
      analogWrite(green, x);
      delay (10);
    }                   //green up/down


    for (x = 255; x > 0; x--) {
      analogWrite(blue, x);
      delay (20);
    }
    for (x = 0; x < 255; x++) {
      analogWrite(blue, x);
      delay (20);
    }                   //blue up/down
    break;
  } //end of action
  x = 0;
}
//////subRoutines
void clearArray() {//subroutine
  for ( x = 0; x < 8; x++) { //clear arrayCall
    arrayCall[x] = 0;
    action = 0;
  }
}
```

Code Listing 3.2. Landscape project

The program uses arrays to store the ASCII code numbers. Arrays consist of a single variable with multiple instances. The number of instances can be declared, which will allocate memory locations. Each of the color arrays is defined in the setup section so that it may be possible to match them later in the main loop. The pin numbers for the LEDs are not random and are used because they can specifically produce Pulse Width Modulation (PWM) in an Arduino Uno. PWM at a specific frequency will vary a square wave pulse width to output a signal that varies in power, as shown in Figure 3.14. (Most modern-day AM broadcast transmitters use PWM as a way to develop amplitude modulation.)

Figure 3.14. A PWM signal increasing in power

The LED landscape lamp we were using sent a ground to light an LED. The power was the common bus, and we stuck with that convention, although it seems a bit backward. After declaring variables, we set the initial power-on condition by grounding the pin for the white LED. In the main loop, The information is received from the Bluetooth transmission and stored as an ASCII number, and converted to a letter to be displayed on the app screen. The pins change to the appropriate logic levels as the user changes colors after entering a subroutine to initialize all LED pins off. Mixing occurs using PWM to varying the LEDs' intensity, which will brighten and then dim before switching to the next color. You may wish to experiment with the timing and produce different effects.

Section 3.6. 433 MHz transmitter and receiver project

You can buy a transmitter and receiver used in this project for less than a cup of coffee, and they have a great many uses. They are so useful, you find them everywhere, but that creates a problem. Communications using these devices are very prone to interference and can cause erratic operation. Make sure the 433 MHz devices are legal to operate in your country. We will run through a quick project to control an LED. If your project does not operate correctly, try moving to a different location. A deserted island would be best. (You can run Arduino Unos with a 9-volt battery snap, so electricity is not necessary.) It may actually help to reduce USB power noise if you use batteries after the boards are programmed. A picture of the transmitter and receiver appears in Figure 3.15. The receiver is pictured on the left, and the transmitter is on the right.

Figure 3.15. 433 MHz receiver and transmitter

You must solder in an antenna after purchasing the modules. There is an open solder pad located next to the small coils on both units. Antennas can be bought or easily made from a thin 17.3 cm 22-gauge solid wire. This amount of wire is one-quarter wavelength at 433 MHz. There are also base loading coil designs that use a longer wire and spiral a coil of about 16 turns starting near one-third of the way up. A little research will show many different antenna designs. You may notice that the breadboards are practically bare. The receiver on the left breadboard has 5 volts and ground coming from the Arduino, and the output is directly connected to Arduino pin 7. An LED and 220 Ohm current limiting resistor connect between digital output pin 8 and ground.

The transmitter is also supplied with power and ground from the other Arduino with its data input directed to pin 7, and we are using Arduino pin 8 as a momentary switch to send a timed pulse. The pins on the modules should be marked, or you can refer to Section 3.3 for a closeup view and a description of the pinouts. The code for this project is shown in Code Listing 3.3 for the transmitter and Code Listing 3.4 for the receiver.

```
//code for 433 MHz transmitter
const int modOut = 7;
const int xmit = 8;
boolean LED1;
```

```
void setup() {
  pinMode (xmit, INPUT_PULLUP);
  pinMode (modOut, OUTPUT);
}

void loop() {
  LED1 = digitalRead (xmit); //button pressed
  delay (40);
  if (LED1 == LOW) {
    digitalWrite (modOut, HIGH);
    delay (100); //100 ms pulse
    digitalWrite (modOut, LOW);
  }
}
```

Code Listing 3.3. 433 MHz transmitter code

After declaring our pin names and variable and defining inputs and outputs in the setup section, the loop checks for the wire from pin 8 to be momentarily grounded; when a ground pulse is detected, it will output a 100-millisecond high-level pulse to the transmitter data pin. The transmitter will send an ASK signal for that amount of time. In Code Listing 3.4. We use a function in the main loop called PulseIn, which measures the pulse width of the transmission.

```
//code for 433 MHz Receiver
const int LED1 = 8;
unsigned long modIn;

void setup() {
  pinMode (7, INPUT);
  pinMode (LED1, OUTPUT);
}

void loop() {
  modIn = pulseIn (7, HIGH); //measures pulse width
  if (modIn > 50000 && modIn < 150000) { //looking for 100 ms, in us.
    digitalWrite (LED1, HIGH);
    delay (2000);
    digitalWrite (LED1, LOW);
```

```
        }
    }
```

Code Listing 3.4.

The PulseIn function is very easy to use but can be somewhat problematic. If you expand on this project, you may wish to use the "millis" timer, as we did in Section Two, Code Listing 2.3. In our "If" conditional section, we are granting quite a bit of tolerance which you can tighten up a bit if there is an interference problem causing erratic operation. The reason for large numbers is that the pulseIn command returns a result in microseconds (us). If all goes well and the transmit pulse is within tolerance, the Receiver LED will light for two seconds.

It is best to start with the transmitter and receiver within a few feet of each other and then separate them to check the range of operation. The battery snaps connected to the Arduino barrel power connector helps to move the units freely. And as was mentioned, if you notice erratic operation, there may be other 433 devices sending transmissions, and you may need to shut them off or change locations.

Chapter Three Summary

There are many low-power devices in operation in the unlicensed ISM bands, many are a part of the Internet of Things, but many are also stand-alone devices like key fobs, TV, appliance remote controls, etc. Equipment in the ISM band is limited to a short transmission range and subjected to interference from other devices using the same frequencies. There are workarounds for minimizing the noisy conditions of these frequencies. Spread spectrum is the most popular technique where frequencies are hopped, transmitting and receiving data over a wide RF range. A new spread spectrum technology called LoRa is also gaining popularity in the Maker Space. It uses a very slow data transmission rate over a very wide band. Frequency hopping can be both active and passive, but precise synchronization must exist between transmitter and receiver in both cases. Bluetooth has a very good range, and under 4.0 standards, transmissions can reach upwards of 100 meters. Wi-Fi was mainly used for wireless LANs but is also used with other devices and is popular for IoT equipment. For extra throughput, sometimes more than one channel is used. All RF transmissions are analog but can be modulated using digital techniques. The three ways to modulate RF are through changes in amplitude, frequency, and phase. Sometimes a combination is used for digital modulation. The devices that we experimented with in this chapter can be used for very complex applications where the controller program usually includes a downloadable library to expand the device's capabilities. In the next chapter, we will begin looking at more powerful transmissions which are licensed and assigned to dedicated frequencies.

Chapter Three Questions

1. Why can't digital signals be transmitted directly using radio frequencies?
2. Give an example where a digital signal can be directly communicated.
3. What type of mode of fiber optics is indexed, causing it to be slightly denser in the center?
4. The name of the mathematical law that explains refraction and reflection in different materials.
5. Give two popular Bluetooth frequencies.
6. How do objects in the path of Bluetooth transmissions affect the range?
7. Explain the process of a material that has a Negative Temperature Coefficient (NTC).
8. What type of temperature coefficient do most materials have?
9. What advantage does a digital device have with a wider bandwidth?
10. What is a disadvantage when wide bandwidth devices are used in a small frequency band?

Chapter 4
Wired and wireless communications

Section 4.1. The telegraph and CW

Smoke signals and other signaling methods were replaced in the mid-1800s with an electrical switching and sounding system called the telegraph. It overtook the most elaborate signaling system of the day, called the Semaphore, also sometimes referred to as the "Napoleonic Internet" or "Napoleonic Telegraph". The system consisted of over 500 maned towers covering approximately 3000 miles across the French countryside. Visual signals allowed for information to be transmitted across the country in a matter of hours and replaced messengers on horseback which previously conveyed the same information over a period of days.

After the electrical telegraph was introduced in 1837 other signaling systems quickly disappeared as the telegraph gained worldwide acceptance. A switch to send current pulses went through wires located along railroad right-of-ways. Stations along the route not only had switches for transmitting the pulses but also an electromagnetic device that produced an audio click when the electricity was detected. Electric current travels through wires at roughly three quarters of the speed of light. The telegraph wiring network was quickly assembled, and cables were even laid under the sea to connect the continents. A modern version of a telegraph switch, called a key, used to make, and break the current flow is shown in Figure 4.1.

Figure 4.1. A code transmit key

The code for the pulses of current was developed by Samuel Morse and contain a series of short pulses called "dots" and longer pulses called "dashes". It is somewhat like the binary bits we use in our digital systems today except that it uses two different characters rather than an on/off scheme to represent two possibilities. The sequence of dots and dashes that make up Morse Code is similar to a digital pulse train and is described in Chart 4.1.

a = • ▬		s = • • •
b = ▬ • • •		t = ▬
c = ▬ • ▬ •		u = • • ▬
d = ▬ • •		v = • • • ▬
e = •		w = • ▬ ▬
f = • • ▬ •		x = ▬ • • ▬
g = ▬ ▬ •		y = ▬ • ▬ ▬
h = • • • •		z = ▬ ▬ • •
I = • •		1 = • ▬ ▬ ▬ ▬
J = • ▬ ▬ ▬		2 = • • ▬ ▬ ▬
k = ▬ • ▬		3 = • • • ▬ ▬
l = • ▬ • •		4 = • • • • ▬
m = ▬ ▬		5 = • • • • •
n = ▬ •		6 = ▬ • • • •
o = ▬ ▬ ▬		7 = ▬ ▬ • • •
p = • ▬ ▬ •		8 = ▬ ▬ ▬ • •
q = ▬ ▬ • ▬		9 = ▬ ▬ ▬ ▬ •
r = • ▬ •		0 = ▬ ▬ ▬ ▬ ▬

Chart 4.1. Morse Code

There is no provision for upper- or lower-case letters in Morse Code as there is for the ASCII code that we examined in the last chapter. In looking at the letters S and O in the chart, we see that the S is comprised of three dots, and the O three dashes. The emergency message of SOS does not stand for "Save our ship" as is commonly believed, but is used because the pattern is very recognizable.

Morse code began to be commonly used on ships at first by flashing lights, and later with a device called a *Wireless* in the very late 1800s. The Wireless transmitted and received the Morse Code pulses on low radio frequencies. Ships at sea could communicate with other ships and with ground-based stations. The low frequencies could travel great distances and the Wireless was not only used for ship communication and navigation, but also to pass messages between passengers and friends and family back at their hometowns. The messages would be received by a ground-based station and delivered via telegram. Most major cities had stations capable of wireless communication. There was even a ground-based station located in the John Wanamaker Department store in Downtown Philadelphia. The Wireless which was developed in 1894 ushered in the modern age of communications.

The very early wireless transmissions used high voltages to generate a spark across a gap. This method of transmission had the disadvantage of creating many spurious signals and consumed a very wide bandwidth. Vacuum tube technology developed in the early 1900s led to the wide use of radiotelegraphy.

Licensed Amateur Radio operators, called Hams, sometimes still use Morse Code because it has the advantage of being able to be heard in poor conditions through harsh interference. As of about fifteen years ago it was an FCC requirement for advanced Amateur Radio license holders but is now a mode only used as a hobby and is referred to as Continuous Wave (CW). The veteran hobbyists can send and receive CW Morse Code messages at over 30 words per minute. There are also programs available where the user interacts with a keyboard and monitor without the need to memorize the code. We will examine such a project later in the chapter.

On RF transmissions of the CW Morse code, the pulses must not have a fast rise or fall time, or spurious emissions will occur called ringing. Rather than sending a squared pulse, the modified pulse similar the one shown in Figure 4.2 can be transmitted.

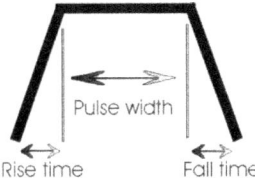

Figure 4.2. CW pulse

The figure over emphasizes the long rise and fall time that is required to avoid key clicks, but the concept is to make the RF output somewhere between a digital style pulse and a sine wave. This is necessary because a square or rectangular wave is made up of the fundamental frequency and harmonic sine waves. Harmonics of square waves are odd multiples of the fundamental frequency. The sample problem given next, is an example.

Problem

Find the first three harmonics of a 100 Hz square wave.

Solution

The first three odd multiples are 3, 5, and 7

F1 = (3)(100) = 300 Hz

F2 = (5)(100) = 500 Hz

F3 = (7)(100) = 700 Hz

The original frequency in the problem is called the fundamental frequency and is the strongest, but the harmonic frequencies will also be present at lower power levels as the multiples increase. You can discover that square waves are built from sine waves by sketching a fundamental sine wave and then overlaying the odd multiples from the same starting point.

Harmonics of RF carrier transmissions that are not modulated with square waves occur at both odd and even multiples of the fundamental frequency. They will also become lower in power as the multiples increase.

Problem

Find the first three harmonics of transmitter with a frequency of 1 MHz.

Solution

The first three multiples are 2, 3, and 4

F1 = (2)(1 MHz) = 2 MHz

F2 = (3)(1 MHz) = 3 MHz

F3 = (4)(1 MHz) = 4 MHz

To reduce the harmonics from being transmitted, a low-pass filter is connected between the final power amplifier and the antenna of a transmitting station. A low-pass filter can be made from passive components such as capacitors and inductors and there a wide variety of designs. The filters' purpose is to pass the low fundamental frequency and attenuate the higher frequency harmonics from causing interference. Our next project demonstrates low pass filtering using audio frequencies.

Section 4.2. Square to sine wave project

It seems almost magical that a square pulse is made of an infinite number of odd harmonics. They are positive odd multiples of the fundamental frequency of the square wave. This helps to explain why Morse code key clicks cause interference during transmission unless the a given slight modification so as not to have sharp squared edges at the front and back side of the pulses. Our project uses a sharp square wave but the circuit we will build will filter frequencies above the fundamental. A slight bit of the fundamental frequency is also attenuated so an amplifier is used to improve the signal strength.

The cutoff frequency of an RC filter circuit is given by the formula $fc = 1/2\pi RC$. The frequency of cutoff is at the 70% point of Vp and any higher frequencies should begin dropping with a gentle roll off. Each instance of a filtering section is called a pole. Normally only a one pole filter is useful because of attenuation even at the wanted frequencies called the pass band. We are using two poles and will amplify the output with an LM386 power amp set up for a gain of 20, where we can calculate (A = 20) by measuring the output over the input. The fc we have for both poles should be equal even though we are using different part values in our RC filtering to obtain the same frequency of cutoff. Filters can be very elaborate, but our demonstration RC filter is the starting point and is known as an L type filter because the resistor and capacitor are in an L configuration (albeit, up-side-down). Sometimes inductors are also used to oppose higher frequencies and are usually placed in series with the signal to be filtered.

Problem

Using the formula given in the above paragraph, find the frequency of cutoff (fc) using an RC filter with a 2.2 k Ohm resistor and a 0.1 uF capacitor?

Solution

$fc = 1/2\pi RC$

$fc = 1/2 \pi (2200) (0.1 \times 10^{-6})$

$fc = 723$ Hz

The 10 uF capacitors placed in series before the LM386 input and after its output to pass AC, but block DC. The Square wave source in the diagram can consist of the Arduino running a tone program made to produce a steady 1 kHz tone by simplifying the program shown in Code listing 1.1, or with a tone generator adjusted to the same levels. We are using between 5 to 9 volts for VCC.

You can build and test the circuit shown in Figure 4.3 by connecting a small speaker to the output and verify it is a cleaner sound than the square wave being produced directly by the Arduino or audio generator. (Be sure to connect a small current limiting resistor greater than 100 Ohms when connecting the speaker directly to the Arduino.) If possible, an oscilloscope can also be used to observe both input and output waveforms. The output should somewhat resemble a sine wave.

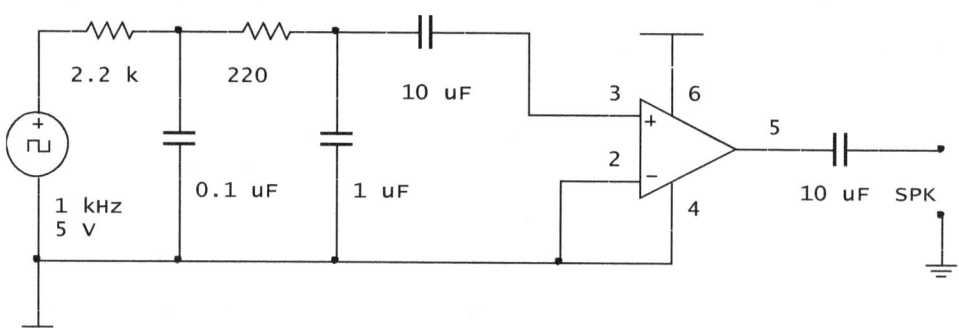

Figure 4.3. Square to Sine Circuit

The fundamental frequency is slightly attenuated since the frequency of cutoff is just below 1000 Hz, but the harmonics are higher and thus will receive more attenuation. The output will not be a perfect sine wave, but nothing is perfect. You might try adding another pole to the filter or change the cutoff frequency to improve the output characteristics. The term cutoff applies to the 70% point (0.707) of the signal, also called the half power point. (0.707 comes up quite a bit in electronics work.)

Section 4.3. Resonance and oscillators

The principle of resonance uses a slight positive feedback to act as *constructive reinforcement* to continue a dampening oscillation. As a continuous wave repeats the process every cycle, but some energy from an input source is required to sustain the condition. If you were to tap a tuning fork, the sound would be noticeable for a while after the impact. Eventually, the sound would *dampen* out, and the fork would require another jolt to continue to make a tone. The effect is depicted in Figure 4.4.

Figure 4.4. Dampening wave

The tone that a tuning fork makes is dependent on its physical structure. Generally, larger objects make lower tones than smaller objects, and this makes sense because we know that lower frequency waves have a longer (larger) wavelength. Resonance also takes place in electronic circuits, where the signal effectively vibrates back-and-forth between two components which produces a high amount of energy exchange. The two electrical components exchange energy in almost the same way as mechanical energy makes a tuning fork vibrate. For an electrical circuit to vibrate at resonance, you need to connect a capacitor to a parallel inductor. The capacitor stores energy in the form of an electric field, and the inductor stores energy in the form of a magnetic field. When the capacitor and inductor are parallel across a source of energy, as in Figure 4.5, it is called a tank circuit.

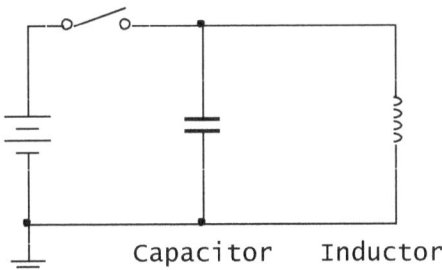

Figure 4.5 Parallel tank circuit

When we close the switch contact, it is like one tap to a tuning fork. The oscillation will quickly dampen out as the circuit's internal resistance transforms the electric current into heat. A small periodic trickle of energy that is in exact phase with the signal is needed to sustain the oscillations.

The physical structure of a tuning fork determines the frequency and wavelength of its tone. In a resonant electronic circuit, the frequency and wavelength are determined by the capacitance and inductance values. The capacitor is given the letter C, and the unit of capacitance is the *Farad*. The inductor is given the letter L, and the unit of inductance is called a *Henry*. A resonant condition exists when each component's resistance to AC called *reactance* is equal. The amount of reactance is expressed in Ohms. Capacitive reactance is given the term X_C, and X_L is used to represent inductive reactance. Reactance is an *imaginary number* because it is perpendicular to the real number line. Real resistance is on the real number line, while X_C is 90 degrees above it, and X_L is 90 degrees below the real number line. The capacitive Ohms and the inductive Ohms are 180 degrees out of phase. No true energy is dissipated by reactance. When $X_C = X_L$, the circuit is in resonance, and the reactive Ohms cancel out and all that is left are the resistive Ohms.

An RF oscillation is a perfect sine wave occurring at the desired frequency, and an oscillator is a circuit that continually produces the waveform. The simplest way of conveying intelligence occurs as coded pulses are sent by turning the transmitter power on and off. The oscillator is the first section of a transmitter, and the switching will be done at a later stage. Oscillators must remain very stable and produce a continuous RF sine wave while the transmitter is powered on. Oscillators are derived from what is termed the *Flywheel Effect*. In a mechanical analogy, a flywheel is somewhat massive, and once rotation begins, its inertia keeps the rotation going with a little reinforcement. In an electrical circuit, the flywheel effect can be produced by the exchange of energy between a capacitor and inductor. Each complete exchange will produce what is essentially one rotation of a 360-degree sine wave cycle. A tube or transistor circuit is used to supply a slight in-phase reinforcement signal for a fraction of each cycle. A capacitor and inductor in series or parallel will have a specific resonant frequency given by the following equation:

$$fr = \frac{1}{2\pi\sqrt{LC}}$$

Where fr is the frequency of resonance, L is the inductance, and C is the capacitance.

Problem

Find the resonant frequency of a signal using a 1 micro-Henry (uH) inductor and a 0.1 micro-Farad (uF) capacitor.

Solution

We use the formula $fr = \frac{1}{2\pi\sqrt{LC}}$

$$fr = \frac{1}{2\pi\sqrt{(0.1 \times 10^{-6})(1 \times 10^{-6})}}$$

fr = 503,292 Hz or roughly 503 kHz

(Just below the AM broadcasting band which starts at 535 kHz.)

The legacy designs for LC oscillator circuits were the Colpitts type which could be recognized by a split capacitor design, and the Hartley type that used a split inductor design. The most common type of oscillator design that is in use today utilizes ceramic resonators or a crystal as the frequency determining device. They operate at a precise frequency utilizing the piezoelectric effect and produce a voltage when physically distorted (bent) and will conversely physically distort as a voltage is applied. A specific resonator or crystal will have a vibrational resonant frequency due to its size and shape. While Crystals have the greatest precision, Ceramic Resonators are widely used as the frequency determining device in most modern oscillators, due to their low-cost and wide availability.

Section 4.4. CW transmitters and receivers

The name continuous wave comes about because the transmitter is oscillating at its RF frequency for the entire time that the Morse Code pulse is being sent. The RF oscillator of a transmitter is the first stage and is usually isolated from the progressive stages by a buffer or driver circuit as outlined in Figure 4.6.

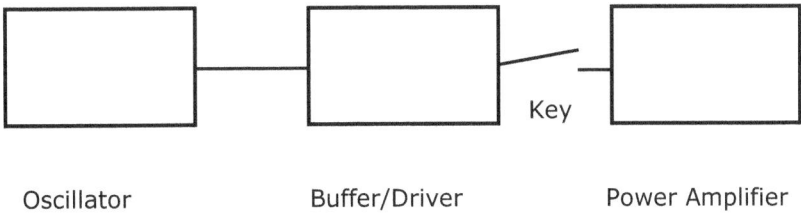

Oscillator   Buffer/Driver   Power Amplifier

Figure 4.6. CW Transmitter

In the block diagram the oscillator is connecter to a high impedance circuit inside the buffer/driver so as not to cause any loading effects. The driver section increases the power to provide input to the power amplifier which may be connected at its output to a low-pass filter and then to an antenna. The key section provides switching to convey coded modulation. The keying circuit should additionally have circuitry to reduce key clicks, or this may be contained at the input to the power amplifier. A RF feedline (not shown) connects to an antenna of a length corresponding to the frequency of the transmission.

The receiver works somewhat in reverse order with an amplifier located first in the chain connected to a feedline coming from an antenna. The amplifier is needed to boost the weak signal since electromagnetic radiation is greatly attenuated with increasing distance and follows the inverse square law shown by the formula:

$$p = \frac{1}{d^2}$$

Where p is power, and d is distance. This formula shows that as the distance between a transmitter and receiver is increased the power at the receiver is greatly decreased.

Problem

The distance between a transmitter and receiver is doubled, what is the resulting decrease in power found at the receiver?

Solution

Using the inverse square law, we have:

$$p = \frac{1}{d^2}$$

$$p = \frac{1}{2^2} = \frac{1}{4}$$

We see that doubling the distance decreased the power at the receiver to one fourth. For very weak signals, such as those found in satellite communication, the receiver has a preamplifier located as close as possible to the antenna, usually placed directly at the feedhorn of a parabolic dish.

For our CW transmitter example, the associated receiver stages would look as shown in Figure 4.7.

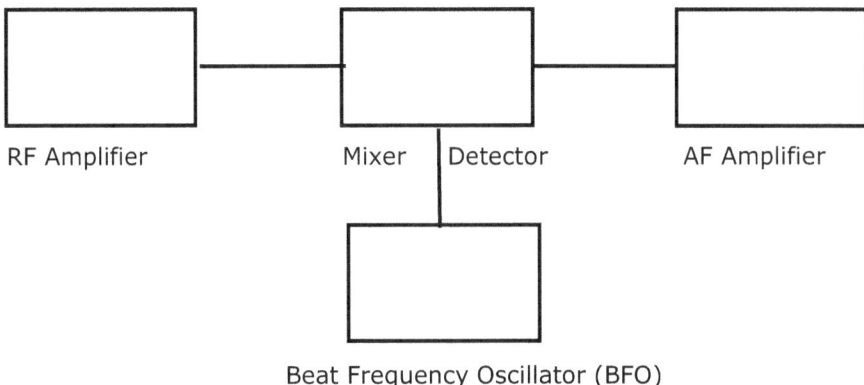

Figure 4.7. Block diagram of a CW receiver

After the transmitted signal is amplified, the detector removes the carrier signal and mixes an input from a local oscillator to produce a tone representing the coded pulses which are than amplified and sent to a speaker. The example that we give is only a simplified representation and more realistic examples will be presented in subsequent chapters.

Section 4.5. I2C communication project

Busses in electronic systems are used to make connections and many different protocols are used for communications. This section examines the I2C bus (Sometimes pronounced I squared C.) The Arduino pins vary with the different model of controller. The pins are shown in chart 4-2 for some of the more popular Arduino boards.

Model	SDA	SCL
Uno	A4	A5
Mega	20	21
Leonardo	2	3
Duo	20	21

Chart 4.2. I2C pins for Arduino boards

An Arduino library contained in the IDE is loaded to simplify the program and only two wires are used to transfer data serially. The data is transferred on the SDA line and the synchronizing clock pulses are sent over the SCL line. The two Arduino boards must also have a common ground to act as a return for both the data and clock lines. The serial monitor screen located in the IDE is used for communication. After connecting the boards and uploading the code, short phrases typed and entered into the transmitter serial monitor will appear on the serial monitor of the receiver. Two computers and two Arduino boards spaced a short distance apart are needed for this project making sure the serial monitor on the IDE is set to "new line" as the program is run. One of the boards is coded as the transmitter and the other as the receiver as shown in Code Listing 4.1 and 4.4.

```
// Wire Slave Transmitter
#include <Wire.h>
int keyBoardByte;

void setup() {
  Serial.begin(9600);
  Wire.begin(8);          // join i2c bus with slave address #8
  Wire.onRequest(request); // go to subroutine function
}
void loop() {
  delay(100);
}
// function executes whenever data is requested by master
void request() {
  if(Serial.available()>0){
    keyBoardByte = Serial.read();
  }
  Wire.write(keyBoardByte); // respond with message
  if (keyBoardByte == 10){ //ASCII 10 is New Line
    keyBoardByte = 0;
  }
  keyBoardByte = 0; //resets the variable
}
```

Code Listing 4.1. Transmitter Code

```
// Wire Master Receiver
```

```
#include <Wire.h>
int x;

void setup() {
  Wire.begin();
  Serial.begin(9600);
}

void loop() {
  Wire.requestFrom(8, 32);    // request 32 bytes from slave device #8
  while (Wire.available()) { // slave may send less than requested
    x = Wire.read();
    if (x > 31 && x < 127){ //Get ASCII text and punctuation
    char c = x;           // receive a byte as character
    Serial.print(c);      // print character
    }
    if (x == 10){         //New line
      Serial.println("");
    }
  }
  delay(100);
}
```

Code Listing 4.2. Receiver code

The I2C bus is an example of coded communication over a wire. The purpose is for synchronized serial data communication and is mainly used to connect a microcontroller board to external sensors over a short distance. A master receiver can connect to multiple slave transmitters on the bus by selecting the desired address. In our code, the receiver sends a request to the transmitter to send any data in the keyboard buffer. The ASCII codes for upper- and lower-case letters, numbers, and punctuation will be displayed on the receivers' serial monitor. The ASCII code 10 is sent when the enter key is pressed and creates a new line on the receiver serial monitor when multiple lines of text are sent by the transmitter. Each line is limited to a length of 32 characters.

Section 4.6. Morse code project

Many people use an audio oscillator in learning to send the dots and dashes that make up the letters and numbers in Morse code. We will run a program on the Arduino that will give us an example of how Morse Code sounds as shown previously in chart 4.1. The program generates the emergency call SOS. Those letters are used because their code is similar and easily recognizable. A small speaker is connected in series to a 120 Ohm resistor from pin 7 to ground. Momentarily grounding Pin 2 will start and stop the Arduino from generating the distress call as shown in Code Listing 4.3.

```
const int spk = 7;
const int onOff = 2;
volatile int toggle;
boolean output;
int dot = 250; //dot time 250 ms;
int dash = 750; //dash is 3 dot times
int space = 750; //space is 3 dot times
int newWord = 1750; //space between words is 7 dot times

void setup() {
  pinMode (spk, OUTPUT);
  pinMode (onOff, INPUT_PULLUP);
  attachInterrupt(0, resetISR, LOW); //hardware interrupt on pin 2
}

void loop() {
  output = digitalRead (onOff);
  delay(150);//debounce
  if (output == 0 && toggle == 0) {
    toggle = 1;
  }

  while (toggle > 0) {
    tone(spk, 2000, dot); //letter S
    delay(dot + dot);
    tone(spk, 2000, dot);
    delay(dot + dot);
```

```
    tone(spk, 2000, dot);
    delay(dot + space);

    tone(spk, 2000, dash); //letter O
    delay(dash + dot);
    tone(spk, 2000, dash);
    delay(dash + dot);
    tone(spk, 2000, dash);
    delay(dash + space);

    tone(spk, 2000, dot); //letter S
    delay(dot + dot);
    tone(spk, 2000, dot);
    delay(dot + dot);
    tone(spk, 2000, dot);
    delay(newWord);
  }
}
void resetISR() { //reset function, shuts off xmitter
  if (toggle == 1) {
    toggle = 0;
  }
}
```

Code Listing 4.3. Morse Code Audio

Everything is based on the time duration of the dot. We are using the code speed of 5 words per minute (WPM) to sound out the letters SOS which consists of the sequence of three dots, three dashes, followed by three final dots. The tone function has three parameters consisting of the output pin which we call "spk", followed by the call for a 2 kHz tone, and the third parameter specifies the time duration of the tone. We define each units' time duration above the setup section in the code. A hardware interrupt is used to stop the transmission since so many delays are used in the program. The program is written for the Arduino Uno where Hardware interrupt 0 is associated with pin 2. If you are using a different board you may need to change the code for the interrupt number and pin. If the code program is running and the interrupt pin is momentary grounded, the Interrupt Service Routine (ISR), located outside of the main loop, will toggle the on/off condition to stop the program.

Chapter Four Summary

The telegraph was the first mode of electronic communication. It utilized a series of two different pulse widths to convey letters and numbers and is somewhat similar to the digital techniques we employ today. It was a device connected by wires but evolved into a wireless system used initially for ship communication and navigation. The magic of electrical resonance when Inductive and capacitive reactance are equal values can tune the wireless transmissions to a narrow frequency at high power and provide for long distance communications. Oscillators use LC tuned circuits or mechanical crystals as the frequency determining device and very little power is required to sustain the oscillations. Computers and microcontrollers use wired connection to transfer data on groups of lines called busses; they also utilize wireless communication such as Bluetooth and Wi-fi.

Chapter Four Questions

1. Give an example of how communications over a distance efficiently took place before the advent of the telegraph.
2. What physical property did a telegraph sounder utilize to reproduce the dots and dashes?
3. Explain the reciprocity of the piezoelectric effect.
4. What is the phase difference between inductive and capacitive reactance at resonance?
5. What are the three lowest harmonics of a 1 kHz square wave?
6. What type of filter can be used on a transmitter to reduce the transmission of unwanted harmonics?
7. Explain the Inverse Square Law.
8. What are a key clicks?
9. What three connections are needed to transfer information on the I2C bus?
10. Explain what is meant by a hardware interrupt.

Chapter 5
Noise

Section 5.1. Electron movement in conductors

To paraphrase nineteenth century French mathematician Lazare Carnot, there is an inherent randomness in every process that tends to dissipate energy in an uncontrollable way. The randomness is referred to by mathematicians and physicists as entropy. It has universal ramifications in fields as diverse as thermodynamics, astrophysics, and biology. Even with this omnipresent force of disorder, it is very hard to generate a truly random number. We will use RF energy to aid in a random number generation project later in the chapter.

Nothing in the universe is perfect. The conductors that we use in electricity and electronics are low in resistance but have a slight amount of resistance due in part to the random vibrations from heat energy of the free electrons impeding the orderly flow of a generated current. The formula for the amount of resistance of a conductor is shown as:

$$R = \rho \frac{L}{A}$$

Where the resistance equals the resistivity of the material ρ (Rho) multiplied by the length of the conductor divided by the cross-sectional area. This means that wires tend to have more resistance as they increase in length, and less resistance as they increase in diameter.

$$\rho \propto \Delta T$$

Resistivity is directly related to the temperature of the material. The general equation for the amount of power of noise cause by thermal vibrations is given as

$$\rho = kT \Delta f$$

Where K is Boltzman's constant, T is temperature and Δf is the bandwidth (BW) of the signal. So, we also see that noise increases as bandwidth expands.

Semiconductors have a negative temperature coefficient, but the temperature coefficient for conductors is positive and is directly related so that as the temperature goes up, the resistance of the conductor also increases. The increase in resistance can be attributed to the ambient heat imparting random vibrational movement in the conductive material. Most of the materials' subatomic particles are tightly locked in place but at any given time there are a number of free electrons. As the temperature increases, so does the resistance, with the random movements of free electrons offering opposition to the orderly electron flow of a directed current.

Along with the increase in resistance as the temperature of a material goes up, the motion of the electrons produces random magnetic fields. The fields are only of concern inside of the conductor since it does not radiate externally due to cancellation - since they are in all directions and out of alignment. As we discussed in Chapter One, the organized movement of electron flow does, however, produce an external magnetic field. The thermal vibration effects within the conductor are given the term Johnson or Nyquest Noise. Both persons were physicists who studied the phenomenon in the early 1920s while working at Bell Labs. The noise appears equally across frequencies and is considered to be a *white noise*. There is extremely low resistance when a conductor is near absolute zero. Work has been underway for decades to create materials with superconductivity at normal temperatures.

Section 5.2. Electron movement in semiconductors

Noise is generated in semiconductors that have a PN junction such as diodes and bipolar transistors. A roadway speedbump is a good analogy of the electron flow through a semiconductor junction. In both instances, energy must be applied to overcome the obstacle. With a speedbump a little extra gas must be applied to raise the vehicle to the top of the obstruction, and then a vibration is felt which produces a noise. In a semiconductor junction an electron must have an electric force (voltage) sufficient to raise is energy level pass through the neutralized barrier. Once it reaches the other side the extra energy is released as photons and electromagnetic radiation. The semiconductor noise can be picked-up on a sensitive radio receiver. In most power supply designs capacitors are placed across rectifier diodes to suppress the electromagnetic interference since it would be appreciable at high currents. This type of noise is referred to as *shot noise*. The term comes from the sound made by the metal BBs of a shotgun shell falling onto a table. (Apparently the physicists first studying this effect were avid hunters.) As is the case with thermal noise, the effects are noticeable across all frequencies. When sensitivity is an issue such as with satellite and other small signal reception, low noise semiconductors can be utilized, and initial amplification can occur at the feedhorn of a parabolic dish to help produce a higher signal to noise ratio. Field effect devices such as JFETs and MOSFETS do not produce very little junction noise.

Section 5.3. External noise

Both previous sections concerned noise generated internal to a device. This section examines noise generated by outside sources. We are again reminded that the term *noise* which we are using refers to unwanted electromagnetic interference. There are three general categories of external noise: human-made, atmospheric, and space noise.

Yon may have heard the low pitch buzzing sound near an electric company substation. The audible sound is made by the large transformer cores vibrating in tune with the 60-cycle current flowing through the coils. (60 cycles is the power line frequency in the U.S. but 50 cycles in some other countries.) There are also inaudible vibrations surrounding the current conducting overhead wires. It is quite noticeable on a cars' radio when driving near high voltage lines while listening to a ball game or talk show on AM. The low buzzing sound on the radios' speaker is again at the power line frequency. This is an obvious source of human made noise but there are many other sources. One of the noisiest devices is ubiquitous in modern society – the computer. One only needs to remove the cover from a PC and hold set an AM radio nearby to hear harsh sounds emanating from the speaker. It is a sharp contrast to the soothing sound of the power line frequency. Computers generate and move data by switching current on and off and use motherboards with many bus lines. The majority of the electromagnetic radiational noise produced is a result of transforming and regulating voltages in the power supply. Almost all modern power supplies use a design called switch mode. The main input voltage is rectified and then chopped at a high frequency before being sent to a small transformer. After rectification of the secondary voltages the operating voltages of +- 12, +-5, and 3.3 volts are obtained. The main chopping frequency is usually varied to provide voltage regulation. The chopper circuit is a notorious producer of noise. Since most other home electronic devices also use switching power supplies, household noise can produce a great deal of interference to RF communications. The external power supplies sometimes called "wall warts" used for charging phones and small appliances are notorious noise producers. The USB output models use switching power supplies and can be very noisy. The larger size models may only contain a transformer and rectifier circuit and don't create as much interference, but they can waist energy by generating a small amount of heat from their transformers if left plugged in when not in use.

Industrial machinery can sometime utilize high voltages and currents which is noise producing along with electromagnetic fields from motors, pumps, and relays. Using shielded Cat-6 cable for Ethernet and other shielded cable for control connections is quite common in manufacturing plants. Medical devices and now even personal portable devices like wristwatches and wearables add signals to the RF spectrum. Human made noise is most detrimental on frequency usage below 500 MHz.

Atmospheric noise such as the booming sound from a thunderstorm can be quite noticeable but when the tremendous surge of current from a bolt of lightning occurs there is also a great amount of RF energy that is emitted. Again, listen to an AM radio during a thunderstorm and you may find it hard to hear anything but noise. Even though there may not be a thunderstorm in your local area, the RF interference from static crashes from across the globe is noticeable on sensitive receiving equipment. Weather conditions not directly associated with thunderstorm activity can also generate RF interference. Atmospheric noise is most problematic on frequencies under 30 MHz.

The solar wind is a constant source of ion bombardment to our planet and produces a steady background noise. This is sometimes exasperated by events which occur on the sun like Coronal Mass Ejections (CME). The dissociated protons and electrons striking our magnetosphere create a tremendous amount of electromagnetic energy which can interfere with both airborne and ground communications up to 3 GHz in frequency. Radio Frequency Interference (RFI) can come from something as beautiful as an aurora as higher energy particles coming from the sun lose energy by giving off photons of light. The solar weather is monitored by the National Oceanic an Atmospheric Administration (NOAA). They release predictions and warnings as conditions warrant. X-ray pulses from the sun can sometimes cause complete radio blackouts on some bands which may last for a number of hours, and can even affect satellite operation.

Cosmic rays are as unpredictable as they are powerful. Supernova, Pulsars, and high energy space events can cause unforeseen soft errors in ground-based computer systems and interfere with worldwide communications. An interesting event occurred during an election in Belgium in 2003 when a recount was triggered after a candidate received many more votes than expected. The computer poling equipment was tested and found to be working correctly but did not agree with the external memory backup. After a thorough analysis, the conclusion was that a bit-flip occurred when a computer memory location was corrupted by a high energy particle collision. It didn't destroy the digital memory location but is thought to have changed a zero to a one, thus generating a much larger number for the candidate than the backup recount showed. Other soft errors have been known to occur periodically in digital systems and may be the result of similar collisions with high energy cosmic particles. It is also believed that these periodic collisions occur with plant and animal cells and can alter DNA, which leads to subtle changes in subsequent generations and causes evolution to occur over time. Our planet is constantly bombarded by powerful unknown sources throughout the universe and even with our planets strong magnetic field, these collisions are constant and since they are random, they are considered to be a source of background noise. The omnipresent background noise of empty space which helped lead to the prevalent thought that the Big Bang was the event which started our universe in motion was found by scientists working for Bell labs in the mid-1960s. They noted a background noise was present on highly sensitive radio receiving equipment using a directional horn antenna, no matter in what open area of the sky the directional antenna was pointed. Afterword, it was found that this residual energy of the universe that surrounds us gives empty space a temperature of just under 3 degrees Kelvin above absolute zero. Nothing in the Universe is perfect or absolute, entropy is a part of life and noise is everywhere.

Section 5.4. Noise specifications

In precise terms, *sensitivity* is the power input required from a received signal to produce an output. Another important term is the *noise floor*. If you consider the analogy of looking around a room and noting the lowest level is the rooms' floor. Even though there may be objects under the floor, they are not visible to you. Similarly, if there is a certain amount of noise present within a given bandwidth at a receiver, and if a desired signal is of less power, than it will go unnoticed. There are some steps that may be taken to lower the noise floor. If the bandwidth can be reduced, then it may be possible to drill below the noise floor. It is good practice to minimize the bandwidth not only to reduce noise but also for efficiency. The noise floor can also be lowered with proper grounding and shielding. Since we reference our signals to earth ground, it is of paramount importance to have a good RF connection to the earth. A single long copper pole of 8 to 10 feet in length inserted into the earth near a communications station is the preferred method to connect to the ground. All communications equipment should be electrically connected to that point with fattened conductors like the tin plated copper stranded grounding strap shown in Figure 5.1.

Figure 5.1. Ground strap

The grounding strap in our photograph is a good conductor and very flexible. As a safety precaution it is not recommended to make any very sharp bends in the event of a lightning strike. All metal communications equipment cases should be connected to the ground strap which then terminates at the ground in the earth. The strap is flattened to allow for the ground currents to have a low impedance path to earth ground. Unlike normal electrical conductors where the resistance of the wire is uniform across its diameter, RF exhibits what is called the *skin effect*, where high frequency currents travel along the surface of the conductor. This effect is due to more inductance being present towards the center of a conductor providing opposition to high frequency current flow due to inductive reactance. This is evident in the inductive reactance formula.

$$Xl = 2\pi FL$$

Xl Ohms increases with both an increase in frequency and an increase in inductance. High power broadcast facilities use long thin sheets of copper flashing to minimize the skin on ground systems. The ground strapping can be seen running along the inside of a radio station antenna tuning box located at the base of an antenna in Figure 5.2.

Figure 5.2. Ground strap at a broadcast transmitter site

Electronic equipment should be enclosed in a metal case connected to a reliable earth ground to minimize interference. This practice keeps signals from entering or exiting the case since the case is at ground potential. In physics, this type of object is known as a *Faraday Cage*. It is good practice to shield high power transmitter circuits so that they won't produce interference. It is also essential to shield receiver front ends to keep interfering signals away from weak signals of interest and allow for good quality amplification. Noise is amplified along with a wanted signal with the same amount of gain. The concept of how noisy of a system we have is given by the signal to noise ratio.

To quantify the amount of noise that is present with a signal we use the *Signal to Noise Ratio* at a specified point in a signal chain. The formula for the signal to noise power ratio is:

$$\frac{S}{N} = \frac{P\ signal}{P\ noise}$$

In log form:

$$\frac{S}{N} = 10 \log \frac{P\ signal}{P\ noise}$$

Problem (S/N ratio)

We find that a radio amplifier stage has 4 milliVolts (mV) of signal level and 0.5 mV of noise. Find the S/N ratio. (Most readings will be with voltage, and an Ohm's law formula for converting to power, P = V² / R is convenient.

Voltage Signal to Noise Ratio : $\dfrac{S}{N} = \dfrac{V^2 \frac{signal}{R}}{V^2 \frac{noise}{R}}$ Note that this is a complex fraction

equal to $\dfrac{S}{N} = \dfrac{V^2\ signal}{V^2\ noise} = \dfrac{4^2}{0.5^2} = 64$

As a log:

$$\dfrac{S}{N} = 10 \log \dfrac{V^2\ signal}{V^2\ noise} = 10 \, \text{Log} \, \dfrac{4^2}{0.5^2} = 18 \text{ dB}$$

You may prefer to use the formula :

$$\dfrac{S}{N} = 20 \log \dfrac{V\ signal}{V\ noise} = 20 \, \text{Log} \, \dfrac{4}{0.5} = 18 \text{ dB}$$

For expressing the loss and gains of noise as signals move throughout different stages of equipment, we use a figure of merit called the Noise Figure. First we find the *noise ratio* which is the signal to noise at the input to the section over the signal to noise at the output.

$$\text{NR} = \dfrac{\frac{S}{N(in)}}{\frac{S}{N(out)}}$$

The noise figure In log form:

$$\text{NF} = 10 \log \text{NR}$$

Problem (noise figure)

In checking between the stages of a radio we have a s/n power ratio of 50 at the input and 5 at the output of a stage. Find the noise figure.

Solution

Step one: Find the noise ratio. $NR = \dfrac{\frac{S}{n(in)}}{\frac{S}{n(out)}} = \dfrac{50}{5} = 10$

Step two: Now find the noise figure. NF = 10 log NR = 10 log 10 = 10 dB

Chapter Five Summary

All conductors have some resistance and inherent noise due to the random motion of free electrons. Just about all conductors have a positive temperature coefficient to resistance and noise. That is to say: as the temperature of the material goes up, the resistance and noise go up. Semiconductors have a negative temperature and conduct current better as temperature increases. (This can cause a vicious cycle if semiconductors like power handling transistors are overheated.) Bipolar semiconductors are inherently noisy as electrons move across PN junctions. Noise from electron movement in conductors and devices are examples of *internal noise*. *External noise* can come from human sources like motors, lights, and electric and data lines. External noise can also come from natural sources like the atmosphere, sun, and the cosmos. Conductive shielding can shunt noise signals to ground and is especially important at the front end of receivers and where other circuits have high gain amplification of weak signals. Shielding is best connected to one common ground point to minimize ground loop currents.

Chapter Five Questions

1. What happens to the resistance of a conductor as the ambient temperature increases?
2. The resistive property of semiconductors has what type of temperature coefficient to heat?
3. Electrons in motion produce what type of field?
4. A shielded box to keep out electromagnetic waves is utilizing the _____ cage effect.
5. Explain the skin effect at high frequencies.
6. Give an example of electromagnetic noise caused by humans.
7. What do the letters RFI stand for?
8. Why is noise produced as semiconductors operate?
9. What happens to the property of inductive reactance as frequency increases?
10. The signal to noise ratio is the signal power divided by _____.

Chapter 6
Broadcast Communications

Section 6.1. Amplitude modulation

Radio communications first began after it was found that an electrical spark could be detected remotely. As described earlier in the text, Morse code was the first "modification" to an RF carrier to provide wireless messages. It can be debated if CW pulses are truly a modulating signal. They might also be compared to digital pulses since they are either present or not. Morse code, however, has two pulse width possibilities: dots and dashes. A digital version of CW, called amplitude shift keying (ASK), was used in our ISM project in Chapter Three.

The first AM broadcast signal was sent from a ground station in Massachusetts to CW operators at sea. It occurred during the Christmas holiday in 1906 when music was transmitted rather than traditional Morse code. Amplitude modulation remained experimental until the first commercial station, KDKA, began broadcasting in 1920 from Pittsburgh, Pennsylvania.

In analog amplitude modulation, mixing an intelligence signal with a carrier signal creates new signals that require a wider frequency bandwidth than CW. The bandwidth of CW may only consist of a few hundred Hertz, whereas amplitude modulated signals need 10 kHz for AM broadcasts.

Modulation is more than just a combination of signals, as shown by our oscilloscope pictures in Figures 6.1 and 6.2.

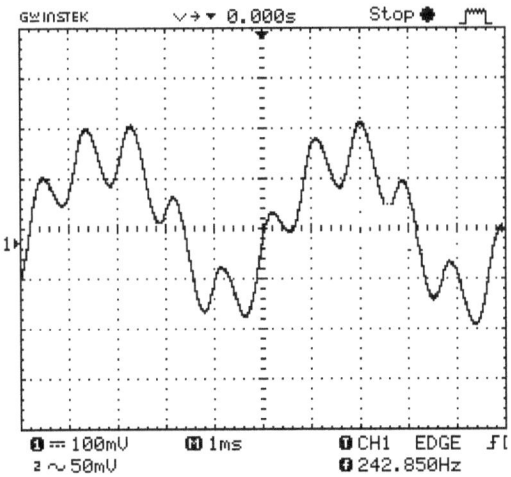

Figure 6.1. Two signals combined

When only combining analog signals, each is present and add and subtract from one another, but signals of a new frequency are not created. In the example shown in Figure 6.1, a higher frequency signal of 1000 Hz is adding and subtracting from a lower but more powerful signal of approximately 200 Hz. Triggering an oscilloscope on such a sample can be very difficult. The type of display in Figure 5.1 may also be seen on an oscilloscope when the probe's ground clip is not making a good ground connection, and 60 Hz combines with the signal of interest.

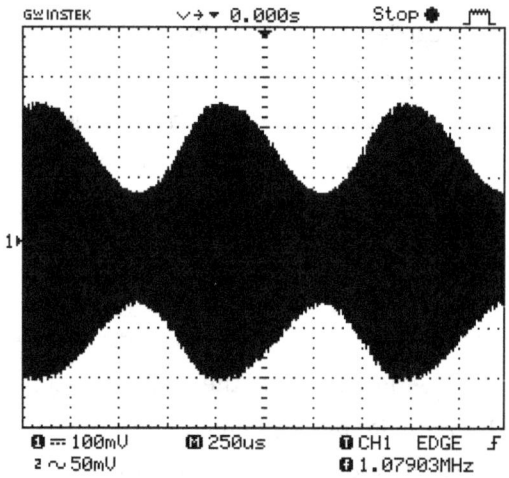

Figure 6.2. Two signals modulated

The amplitude modulated signal shown in Figure 6.2 is a 1000 Hz signal *mixing* with a 1 MHz carrier. Because of the much higher frequency, the carrier appears blurred on the display. Figure 6.3 shows the concept of amplitude modulation more plainly.

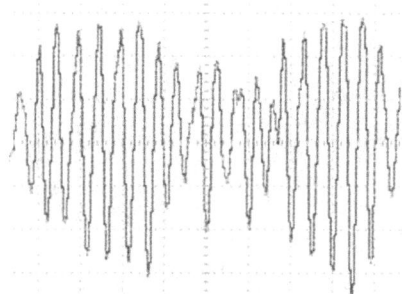

Figure 6.3 AM Modulation concept

For mixing to occur, both signals must be input to a nonlinear device such as a diode, tube, or transistor. The nonlinear β ac curve for a common emitter NPN bipolar junction transistor (BJT) circuit is illustrated in Figure 6.4 and shows that increasing base current results in increasing collector current.

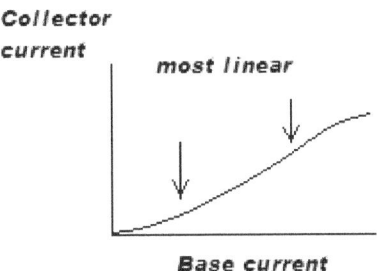

Figure 6.4. Transistor curve

If base current is near zero, the collector current is also near zero (called cutoff). As the base current is increased, the most linear region is presented and would be the preferable area for something like audio amplification. As the base current increases, the collector current approaches saturation – where the transistor behaves like a switch in the on position, fully conducting from collector to emitter. The nonlinear areas are nearest to the current's cutoff and saturation areas. These areas provide the most distortion and are where modulating can best occur.

The amount of mixing can be determined using an oscilloscope display like that of figure 6.2 or 6.3 using the formula shown:

$$\text{Percent of modulation} = \frac{V_{max} - V_{min}}{V_{max} + V_{min}}(100)$$

Problem

Find the percent of modulation of an AM radio signal with a maximum peak voltage of 2 volts, and a V min of 1 volt.

Solution

Substituting the given values into the formula:

$$\% \text{ Mod} = \frac{V_{max} - V_{min}}{V_{max} + V_{min}}(100)$$

$$\% \text{ Mod} = \frac{2-1}{2+1}(100)$$

$$\% \text{ Mod} = \frac{1}{3}(100) = 33\%$$

100% modulation of an AM signal is full loudness, and 0% modulation produces no sound. In our problem, 33% would be a low volume. Care must be taken not to exceed 100% modulation, or the signal will cause distortion and interference. This is especially true for the negative peaks. The best way to find the percent of modulation is to use a single test tone, not music or voice. Music and voice signals are too complex to examine on most test equipment. A standard tone frequency used for test purposes is 1 kHz.

The mixing process, also called heterodyning, outputs the initial fundamental frequencies and produces the sum and difference of the frequencies. This process is used for modulation in the transmitter and demodulation in some receivers. If a single tone mixes with the carrier, the new sum and difference frequencies produced are called side tones, and would look like Figure 6.5 on a piece of equipment similar to an oscilloscope, called a spectrum analyzer.

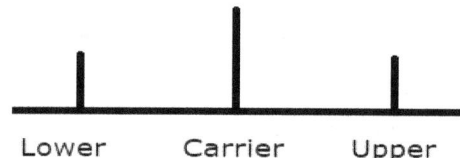

Figure 6.5. Modulated AM frequencies

Frequency is on the x-axis in the figure, and power (voltage) is on the y-axis. At 100 % modulation, each sidetone may contain up to as much as 25% of the total power. The frequency of the tone determines spacing above and below the carrier. With more than one frequency modulated, the areas above and below the carrier would be called sidebands containing two groups, or bands, of frequencies appearing somewhat rectangularly on a spectrum analyzer.

The entire transmitted wave shown on the oscilloscope is called an *envelope*. You can almost imagine the sound portion of the wave in Figure 5.2, riding on top of the higher frequency carrier wave, with its reflection on the bottom. Each contains identical information, and on voice communications systems, such as Ham radio, only one sideband needs to be transmitted and would require much less spectrum. This mode of operation is called Single Side Band (SSB). The carrier is removed with a balanced modulator, and the unneeded sideband is filtered. The carrier is reinserted at the receiver, and the signal is then demodulated.

AM Broadcast stations transmit the entire envelope for best audio reproduction, although the sound fidelity is restricted to 5 kHz by the Federal Communications Commission. The audio frequency limit allows AM stations to occupy the regulated channel spacing of 10 kHz. The entire AM broadcast band runs from 540 kHz to 1700 kHz. Figure 6.6 shows a backup AM transmitter site located remotely from the broadcast studio.

Figure 6.6 Remote backup transmitter

The transmitter is on the right side of the picture and consists of an older model Class C type transmitter. The studio audio is received by a microwave link with the receiver at the top of the equipment rack. The equipment under the link contains modulation meters and the station's local control unit. Normal operation of the equipment is done remotely from the studios.

When modulation is performed in the last stage of a transmitter, it is called high-level because the power of the modulating signal must be brought up to meet the high power in the RF final power amplifier stage. Any amplifiers located after modulation has occurred must be linear to provide an undistorted output, such as in the low-level style. Figures 6.7 and 6.8 show simplified block diagrams for both high- and low-level transmitters.

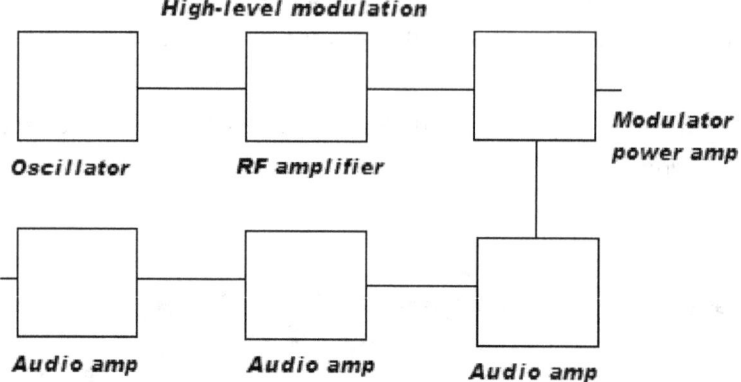

Figure 6.7. High-level modulation

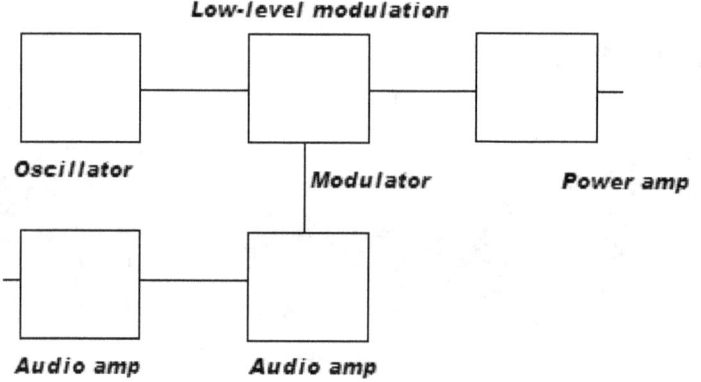

Figure 6.8. Low-level modulation

Section 6.2. Classes of amplifiers

An amplifier's class is determined by the amount of time per cycle the device draws power. From the transistor curve shown earlier in the chapter in Figure 6.4, Class A amplifying devices are biased to run near the center of the curve, where they conduct current 100 percent of the time and draw current even without an audio input signal. Class A provides high-quality amplification and is used in transmitters' front-end stages. Class A amplifiers are also used in audio equipment front-end stages, with class AB providing current to run speaker systems. Class C devices are biased to run in the nonlinear sections of the amplifier curve, where they draw power for only a fraction of the cycle and rely on the fly-wheel effect of resonant circuits.

Class C RF amplifiers are very efficient since they draw current on only a small wave section. They also produce harmonics and require a low-pass filter to be integrated before the signal reaches the antenna near the output of the high-level modulator power amp stage shown in Figure 6.7. The class C amplifier uses the resonance effect discussed in Section 4.2. In low-level modulation, like that of Figure 6.8, any amplifiers located after the modulator must be biased toward the linear center of the transistor curve and run as Class A to avoid distortion. We have included sections in the Appendix that help to understand how amplifiers work.

Electric bills can be quite high to supply the energy needed to operate a broadcast transmitter, and Classes D and E are being used for better efficiency. The remote facility we featured in the last section also has a newer Class D pulse width modulation (PWM) 10,000 Watt transmitter shown in Figure 6.7.

Figure 6.7 AM class D transmitter

PWM produces rectangularly shaped pulses where the widest pulses represent the most positive level of a modulating sine wave, and the narrowest pulse widths represent the most negative peaks. Figure 6.8 is a drawing of a PWM output.

Figure 6.8. Sine wave to PWM

PWM pulses are sent to a reactive tank circuit where a normal amplitude modulated signal is developed. The transmitter in Figure 6.7 uses lower voltages but at higher currents. Many of the power amplifier circuit boards are connected in a parallel configuration and are hot-swappable for ease of maintenance.

Even with the efficiency of class C and D amplifiers, the audio stages are always linear to avoid distortion. Broadcast AM sound quality is limited by a 10 kHz channel width and only 5 kHz of audio, but with a good transmit audio chain and a good receiver, music on AM is listenable. AM radio was the most popular broadcast medium until low-cost FM receivers began to proliferate in the late 1970s. AM transmissions are also more prone to noise and other interference, and AM stations today mainly feature sports and talk shows. AM stereo was made available but isn't used due to interference problems that mainly occur after dark – when the transmitted frequencies of AM broadcasts can travel great distances due to ionospheric reflections. Digitalization of the audio is also not possible because of the limited bandwidth. These drawbacks are overcome with FM broadcasting which we will examine after a project.

Section 6.3. Pulse width modulation project

This project demonstrates PWM. Not all pins on an Arduino are capable of producing PWM (check your models' specifications.) We will modify a snippet of the code from Section 2-5 from our outdoor lighting project to use PWM to fade an LED. We are connecting the LED and 220 Ohm series current limiting resistor between Arduino pin 3 and ground, and uploading Code Listing 6.1 to produce a brightening and dimming illumination.

```
const int LED = 3; //pin 3 is one of the PWM pins
int x = 0;
void setup() {
  pinMode (LED, OUTPUT);
}
void loop() {
  for (x = 0; x < 255; x++) { //LED up
    analogWrite(LED, x);
    delay (4);
  }
  for (x = 255; x > 0; x--) { //LED down
    analogWrite(LED, x);
    delay (4);
  }
  x = 0;
  analogWrite(LED, x);
  delay (100);
}
```

Code listing 6.1. Lighting an LED with PWM

The two *for loops* in the code raise and lower the pulse width and consequently raise and lower the brightness of the LED. This project works best if an oscilloscope is available to connect to pin 3 while the LED is changing brightness. The changing PWM pulses from the Arduino can be observed on an oscilloscope to be of short pulse duration when the LED is dim, and widened to produce brighter intensity. More power is delivered to the LED when the pulse width is greatest and causes the LED to produce the greatest amount of light. In an RF application, the transmitted signal would follow the same scenario after being converted to sine waves.

Section 6.4. Frequency modulation

Frequency modulation is used for most two-way radios, and many public service operations are also now digital. These communication systems use narrow-band FM, where the audio quality is similar to AM radio but without interference issues. The channel size of broadcast FM is much wider than AM to allow for high-quality sound. A maximum +- 75 kHz deviation from the center frequency carrier is used within an FCC channel authorization of 200 kHz. All allocated FM radio channels are odd in number.

The rate and the amount of deviation are two important FM concepts. The rate of deviation is proportional to the audio frequency, and the amount of deviation is proportional to the loudness of the audio. An example from an oscilloscope appears in figure 6.9, where one audio wave, from one positive peak to the next, causes a corresponding carrier deviation. A children's toy called a slinky - made from a spring, is analogous to the look of a modulated FM carrier wave.

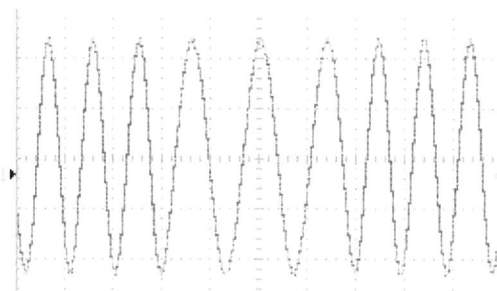

Figure 6.9. FM carrier deviation

Unlike AM, where there are two sidebands, theoretically, FM deviation produces an infinite number of sidebands; however, the power drops significantly as the frequency moves away from the carrier. Sideband formation can be found mathematically through a process called Bessel Functions. A good approximation for the required bandwidth may be intuitively calculated with Carson's Rule using the following formula:

$$BW = 2 (\Delta F + F\ max)$$

(Where the bandwidth is found by doubling the quantity of the maximum frequency deviation plus the highest audio frequency.)

Problem
Find the bandwidth for a mono broadcast FM station where the highest quality audio is 15 kHz.

Solution

Since the maximum deviation is 75 kHz and the highest audio frequency permitted is 15 kHz, we use Carson's Rule:

$$BW = 2\,(75{,}000 + 15{,}000)$$
$$BW = 180\text{ kHz}$$

The vast majority of stations provide two-channel stereo broadcasts through a process for multiplexing analog audio, and some include information for Radio Data Systems (RDS), which provide digital text information to RDS-enabled receivers. Some stations also offer additional compressed analog and digital sub-channels, which would take up even greater bandwidth if the modulation index wasn't reduced. Since the FCC channel spacing is 200 kHz, stations in close geographic proximity are assigned channels spread out across the band to provide frequency separation to avoid interference. Unlike Amplitude Modulation, FM is far less prone to interference from external noise since amplitude spikes and fluctuations are ignored in the receiver's amplitude limiter circuitry.

FM transmitters can fluctuate the carrier's frequency with a reactance modulator. This is a tuned resonant circuit with variable reactance. Commonly the capacitance is varied with a particular diode called a Varactor, as shown in Figure 6.10.

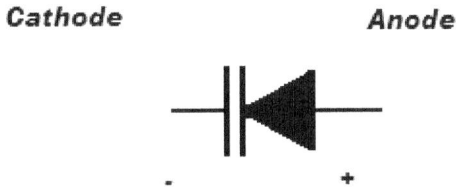

Figure 6.10. Varactor diode

A varactor diode uses the usually unwanted characteristic of capacitance found between the PN junction sections. The diode's capacitance can be varied using a reverse DC voltage placed across the device. As the reverse DC voltage is increased, the junction widens, and the capacitance is reduced. Fluctuating the amount of capacitance in a varactor placed across an LC resonant tank circuit can vary the frequency to correspond to the amount of needed deviation. This is an example of direct frequency generation. (Indirect FM generation uses phase modulation, which is multiplied.) Varactors are also utilized in electronic tuners of receiving equipment. Another method for frequency modulation is to use a Voltage Controlled Oscillator (VCO), where a control voltage is responsible for changing the modulating frequency deviation.

Section 6.5. Television

Great efforts have always been made whenever enhancements modify existing broadcasting services to ensure backward compatibility. When television went from black-and-white to color broadcasting, popularized in the 1960s, old TV receivers continued to work, and upgraded models provided a color picture. A complete switch-over from analog to digital full-power television in the U.S. took place in 2009, and all television receivers needed to be replaced or use an external converter. Analog broadcasting used AM modulation for the picture and FM for the sound on a 6 MHz bandwidth channel. (Keeping in mind that AM radio channels are 10 kHz – quite a difference!) Video contains a large amount of information and consequently requires large bandwidth. Efficient utilization of TV spectrum is achieved using Vestigial Sideband transmission. It is similar to Single Side Band (SSB), used for two-way radio voice communication, except that the carrier and a small part of the other sideband are also transmitted. The 6 MHz bandwidth channel is still used for digital broadcasting. The video is highly compressed, and redundancies are removed for the encoded data to fit into the channel. MPEG-2 compression and Vestigial modulation of U.S. digital broadcasts provide 19.39 Megabits per second (Mbps), which can be divided to broadcast subchannels of lesser quality. Video uses wideband amplifiers since each channel occupies 6 MHz (600 times wider bandwidth than an AM broadcast radio station.)

Over-the-air TV broadcasting takes a great deal of spectrum, and local channels are well-spaced to avoid interference. VHF and UHF frequencies are used with VHF split between 54 to 88 MHz and 174 to 216 MHz. UHF frequencies hold the higher channels from 470 to 700 MHz. The UHF TV band went up to 885 MHz before analog TV was switched over to digital. Some channels were moved to different frequencies, and the then available frequencies between 700 and 885 MHz were auctioned to cell phone companies by the FCC.

As the frequencies go up, the RF transmissions take on microwave properties. Special equipment, such as waveguides rather than coax, and specialized power tubes like magnetrons and klystrons must be used, as pictured in Figure 6.11.

The waveguide can be seen vertically to the right of the tube. Waveguides are hollow pipes of a specific diameter that transfer RF waves and are used in place of coax to connect to antennas.

Figure 6.11. TV transmitter power tube and waveguide

If you have ever cooked food or warmed up coffee in a microwave oven, you have used a waveguide. Inside a microwave oven, a magnetron tube is an oscillator and power amplifier, and the waveguide conducts the energy into the cooking cavity. Like fiber optics, waveguides have total internal reflection and are an extremely low-loss transmission line. However, they are tough to work with since they are essentially hollow copper pipes. The next level down is hardline pictured in Figure 6.12.

Figure 6.12 Coax and hardline

The hardline in Figure 6.12 is pictured to the right of standard RG-8/U, flexible low loss communications coax. Not quite as tough to work with as waveguides, hardline is also made of copper, and it is covered with an airtight insulated exterior. Whereas a waveguide has no center conductor, a hardline has a smaller centered internal pipe positioned with plastic standoffs along its length. The center conductor is hollow because at high frequencies, due to the skin effect, electrons only travel along the surface of a conductor. The line is often pressurized with nitrogen to eliminate arcing at high voltages between the center conductor and shield. Hardline is very low loss and is used with many FM broadcast transmitters and some TV facilities.

Chapter Six Summary

AM mixing (heterodyning) information with a higher frequency carrier signal produces an upper and lower sideband containing identical information. The bandwidth is equal to twice the highest modulating frequency. The percentage of modulation can be calculated and must not be greater than 100 %, or distortion and spurious emissions will interfere with other stations. Frequency Modulation (FM) varies the carrier frequency above and below the center frequency. The rate and the amount of deviation are two important FM concepts, where the rate of deviation is proportional to the frequency of the audio, and the amount of deviation is proportional to the loudness of the audio. Interference will result if the deviation exceeds 100%. Carson's Rule is a convenient way to solve for FM bandwidth. Television uses wideband amplifiers since a TV channel bandwidth is 6 MHz. Vestigial modulation transmits the carrier, a single sideband, and a vestige of the other. Television channels reach 700 MHz, and unique microwave components must sometimes be used, including waveguides to replace coax.

Chapter Six Questions

1. What are three outcomes of mixing a 1 kHz tone with a 1 MHz carrier using Amplitude Modulation?

2. Over what percentage of AM modulation will distortion occur?

3. What would happen if Broadcast AM audio is not filtered and sound over 5 kHz is modulated and transmitted?

4. Explain Pulse Width Modulation.

5. How does loud audio affect a Frequency Modulated signal?

6. What is the rule for finding the bandwidth of an FM transmission?

7. What type of diode can be used as a variable capacitor?

8. FM can be produced in a _____ modulator.

9. Television equipment uses what type of bandwidth amplifiers?

10. Television at higher frequencies may use what type of feed line, rather than coax?

Chapter 7
Broadcast receivers

Section 7.1. AM receivers

The earliest receivers to demodulate AM radio broadcasts were called *crystal radios*. The crystal working in conjunction with a thin piece of metal called a *cat's whisker* works as a rectifier. In the last chapter, we saw an oscilloscope print-out showing what appears to be the sideband audio riding the positive and negative sides of the carrier wave. The crystal acts as a rectifier and removes the bottom part of the envelope. A small capacitor placed in parallel from signal to ground eliminates the remaining high-frequency carrier, leaving the lower frequency audio wave seen on the top half of the envelope. We will demonstrate the actions of receiving an AM signal and detecting the audio in the next few pictures. The entire AM modulated transmit envelope is shown on an analog oscilloscope in Figure 7.1

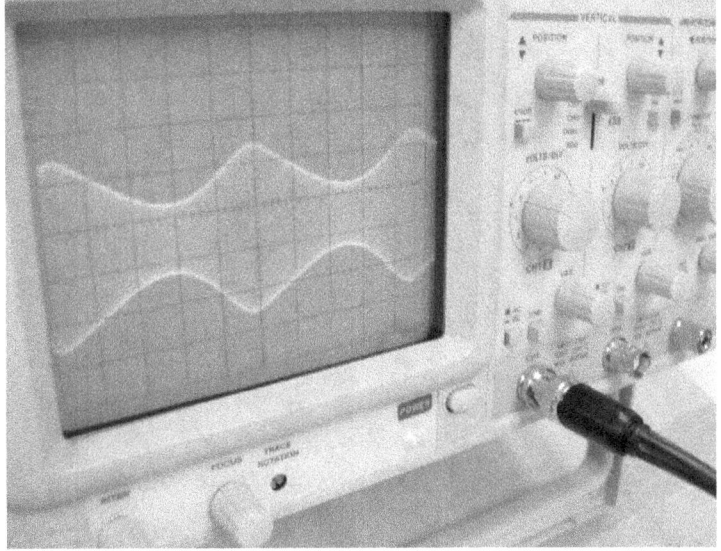

Figure 7.1. AM envelope

The cat's whisker and crystal are replaced with a diode shown as part of the circuit in Figure 7.2. Usually, a diode made of germanium would work best because it has a lower forward voltage drop than silicon (0.3 vs. 0.7 Volts.) The transformer can connect to a resonant tuning circuit and a long wire antenna.

111

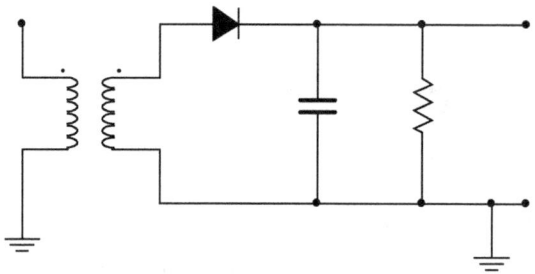

Figure 7.2. AM Demodulating circuit

The capacitor shunts the high RF frequency carrier to ground after rectification occurs. The audio wave is developed across the resistor, as shown on the oscilloscope in Figure 7.3. The output is the sound broadcast from the AM radio station. We used a 1 kHz tone to modulate the carrier in our example.

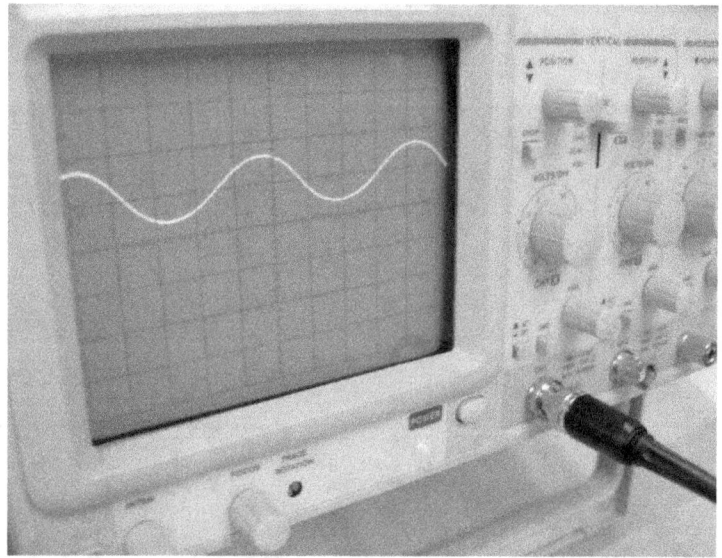

Figure 7.3. Recovered audio

There is usually no amplifier in a crystal radio set, and headphones must be used to hear the sound. However, the receiver requires no power source. Connecting a powered audio amplifier to the circuit would produce sound similar to a typical AM radio receiver. The section just discussed is called a *detector* because it detects the audio. It can also be called a demodulator.

A transmitting radio station's frequency must be selected in a receiver. One method of accomplishing this is to feed the antenna signal across a tunable tank circuit. The listener can choose the frequency of the desired radio station by adjusting the resonance of a parallel tank circuit consisting of capacitance and inductance. It is possible to use a variable capacitor or inductor as the tuning component. Like the one pictured in Figure 7.4, crystal radios have been around a very long time and are Tuned Radio Frequency (TRF) devices.

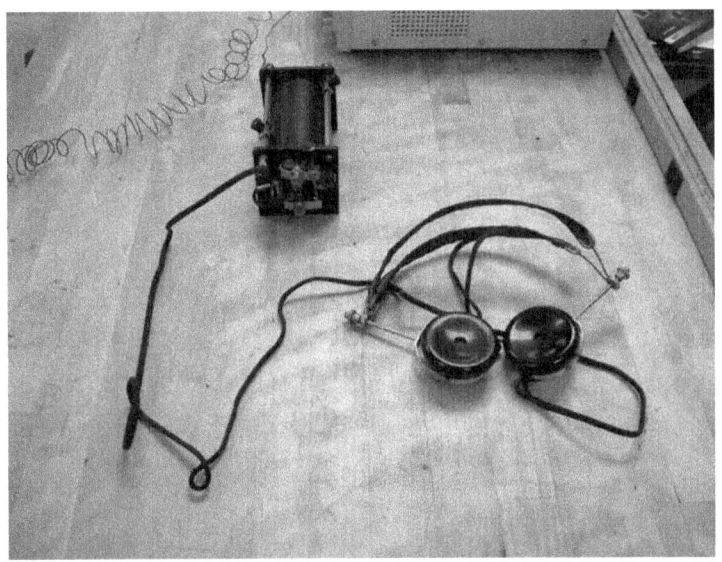

Figure 7.4. Actual 1930s-era crystal radio

Selectivity is the ability to tune in a single station without receiving interference from others, while *Sensitivity* allows for weak signals to be received. S*uperheterodyne* receivers provide a vast improvement over a TRF radio in both selectivity and especially sensitivity. They are the radio receiver design commonly used today. Just as the audio is mixed with the carrier on a modulated transmission, the superheterodyne receiver mixes a locally generated tuning frequency to develop a fixed intermediate frequency (IF). In standard AM broadcast radios, the IF is 455 kHz, and 10.7 MHz is used for FM radios. When mixing occurs, a sum and difference frequency is produced just as in a transmitter. Those signals and the two original frequencies are sent to high Q tank circuits resonating to the desired IF frequency. We usually want to lower the frequency and eventually have audio, so we use the difference frequency between the radio station and the *local oscillator* (LO). IF stages have high-gain narrow-band (high Q) amplifier circuits to boost the signal before it reaches the detector. The demodulated signal is then fed to an audio amplifier and speaker. Figure 7.5 is the block diagram of a single conversion superheterodyne (superhet) receiver.

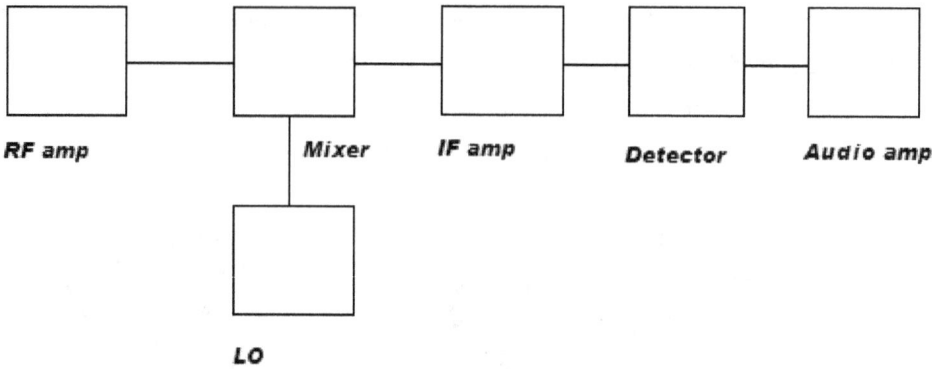

Figure 7.5. Superheterodyne receiver block diagram

Problem

Determine the frequency of the Local Oscillator in a standard AM broadcast receiver if a station at 570 kHz is received. Also, find the image frequency that is produced.

Solution
Part a:

Since we are using the difference between the LO and incoming signal to produce an IF of 455 kHz:

LO − 570 kHz = 455 kHz
LO = 570 kHz + 455 kHz = 1025 kHz

(The LO tuned to 1025 kHz gives us the 570 kHz station since 1025 − 570 = 455.)

The IF amps are resonant to the 455 kHz IF frequency; however, another unwanted input RF signal called an *image frequency* can mix with the local oscillator and produce the 455 kHz, as shown in part b of the example:

Part b:
X − 1025 kHz = 455 kHz
X = 1025 + 455 = 1480 kHz

(The LO tuned to 1025 kHz also gives us an unwanted station at 1480 kHz, since 1480 − 1025 also equals 455.)

The image frequency is within the AM broadcast band and would cause interference to our receiver; however, the tunable tank circuit on the receiver's front end is only resonant to the selected broadcast frequency and filters out the image frequency. For even better reception, two LOs and separate IF stages are used to maximize selectivity and sensitivity. It is known as dual conversion.

Automatic Gain Control (AGC) keeps the strength of incoming RF signals equal so that a nearby transmitter doesn't sound any louder than a distant transmitter. AGC voltage is developed in the detector stage and connected to the amplifiers towards the receiver's front end. Higher AGC voltage reduces the amplifier's gain.

Section 7.2. FM receivers

The two general types of modulation are AM and Angle. Angle modulation consists of varying either the frequency or the phase. Phase modulation slightly shifts the sinewave without completely changing frequency. (Phase modulation is widely used in digital modes of communication.) As we noted in the last chapter, FM transmitters have a center carrier frequency, and modulation causes a higher and lower deviation in frequency. Sound quality is enhanced in FM broadcasting using *pre-emphasis* in the transmitter to boost the loudness of higher audio frequencies, which are then brought back to normal in the receiver using a *deemphasis* circuit. This helps maintain proper audio response since highs are attenuated a bit with FM.

FM receivers use an entirely different type of detection stage, where frequency deviation is interpreted as audio. One of the easiest ways to do this is with a *Phase Lock Loop* (PLL). A PLL is a versatile circuit that can keep from drifting off frequency by generating an error voltage to correct the drift. In a receiver, the error voltage follows the shifting modulated carrier of an FM transmitter. The error voltage resembles the audio wave. With no sound, the transmitter would output the center carrier frequency only. Analog broadcast FM is capable of reproducing audio up to about 15 kHz. Digital broadcasts can produce CD-quality sound up to 20 kHz. There may also be more than one program on a digital broadcast, along with digital information such as the artist's name and song title. Additional decoding circuitry is incorporated into a digital radio receiver.

Along with AGC, mentioned in the last section, limiter circuits are employed in FM receivers to cut off any amplitude fluctuations which may be caused by external noise. The *capture effect* also helps to reduce noise and maintain selectivity.

It causes the receiver to lock on to the more powerful FM signal if a distant or adjacent station may become active. The block diagram for an FM receiver is similar to Figure 7.5, but with the detector called a discriminator. There is also a stereo demodulator section. A limiter stage clips noise spikes like those caused by lightning and other interference that would be problematic to amplitude modulation. Radio noise is called Radio Frequency Interference (RFI), and if it is caused be electric wiring and devices like motors it is termed Electromagnetic Interference (EMI).

Chapter Seven Summary

The most basic AM receiver is a Tuned Radio Frequency (TRF) device. A superheterodyne receiver contains an Intermediate Frequency (IF) amplification stage preceded by a mixer. The mixer has two inputs; one is the tuned and amplified RF from the radio station, and the other RF input is from a local oscillator. As the LO is tuned to different frequencies, individual broadcasting station's RF signals are converted to the intermediate frequency. A second IF of lower fixed frequency can be added for even more sensitivity. This includes a second mixer and LO. Having two different IF sections is known as dual conversion. The stage in an AM receiver where the audio demodulation occurs is called the detector. In an FM receiver, the detector is called a discriminator. Automatic Gain Control (AGC) voltage is developed in the demodulator stage and used to reduce the gain of the RF and IF amplifiers. FM receivers include a limiter to eliminate amplitude peaks which could cause noise from external sources. Since high-frequency audio tends to be slightly attenuated on the transmit side, the high frequencies are boosted by a pre-emphasis circuit. FM receivers have a deemphasis circuit to normalize the sound. The capture effect also helps to lock onto a transmitting station and reduce interference from nearby transmitters.

Chapter Seven Questions

1. When looking at an AM envelop of an audio test signal on an oscilloscope, what is the high frequency wave between the peaks and valleys?

2. What is the name of an AM receiver without an IF section?

3. Being able to specifically tune in one frequency and ignore others is an example of high _____.

4. A standard IF frequency for AM receivers is _____.

5. Explain AGC.

6. What is the frequency of an LO if a station transmitting on 790 kHz was being received using an IF of 455 kHz?

7. How is the IF section changed if a different radio station broadcasting on a higher frequency is selected?

8. What are the two types of angle modulation?

9. What is the purpose of a capacitor following a diode in a simple AM detector?

10. What would the frequency deviation of an FM station broadcasting at 102.9 MHz be if no audio was being transmitted?

Chapter 8
Antennas

Section 8.1. Vertical antennas

A transducer is a device that converts energy from one form into another. A microphone converts acoustic mechanical energy into electrical energy as a voltage. An earbud or speaker converts electric current into mechanical energy that produces acoustic waves. An antenna is also a transducer. As electric current flows back-and-forth in a transmitter's antenna conductor, an electromagnetic field is produced, propagating perpendicularly around the conductor. The electromagnetic field travels outward into the surrounding space at the speed of light as a transverse wave. This process was described in section 1.1.

Current is produced within an antenna conductor if electromagnetic waves traveling through space cut across the conductor. Similarly, magnetic induction is produced by transformer interaction between the primary and secondary coil windings. Magnetically permeable cores are sometimes used to enhance the process at lower frequencies. Hysteresis effects become more pronounced at higher frequencies, and energy is lost as heat produced in the core. Antennas work best if they are high off the ground and kept away from other conductive objects unless directivity is desired. Transformers and inductors are coiled to maximize the magnetic fields, whereas antennas are best kept somewhat straight for the best propagation of electromagnetic radiation. Coiled antenna sections are sometimes used to reduce the overall antenna size.

Antenna length should match the wavelength of the frequency of the electromagnetic energy. Antennas can become quite large at low frequencies. The formula for electromagnetic wavelength (λ) is:

$$\lambda = \frac{c}{f}$$

Where λ (lambda) is the wavelength, c is the speed of light, and f is frequency.

Problem

Find the height in feet for a vertical antenna for an AM broadcast station at 570 kHz.

Solution

Rounding off the speed of light we have 186,000 miles/sec or (186,000)(5280) = 982,080,000 feet/sec.

$$\lambda = \frac{982,080,000}{570,000} = 1,723 \text{ feet}$$

A fraction of the wavelength may be used with a slight reduction in antenna gain. Many AM broadcast stations use a half or quarter wavelength. In our problem, we have 861 feet for a half wavelength and 431 for a quarter wavelength antenna, which is more manageable. Capacitive hats can also be installed at the top of the structure to reduce the size further, but again there is a slight reduction in gain. FAA regulations require a lamp on antennas higher than 200 feet. You may notice that most cellphone towers are just below this limit, so they need not be illuminated at night. Using a full wavelength on a vertical antenna is somewhat counterproductive because the Earth acts as a reflector and doubles the effective size of the antenna. Ground radials buried a few inches perpendicularly circling the tower are used to enhance the ground reflectivity. They consist of a number of conductors grounded at one point, radiating outward from the base of the antenna, as drawn in Figure 8.1. Radial length is best if they match the antenna length; the more radials, the better.

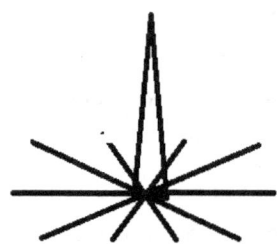

Figure 8.1. Ground radials around a vertical antenna

The transmitter in our example problem would be connected through coax to the base of the tower, which rests on a sturdy insulator. The conductor connected to the transmitter's output connects to the tower, while the shield of the coax is connected to the ground radials. Since the entire tower is radiating and electrically floating from ground, an open ground rod will allow lightning to jump a small gap if the tower is struck. The FCC regards AM transmitter output to be the licensed power. The frequencies in the AM broadcast band are considered to be medium waves.

Medium waves can reflect off the ionosphere at night, and many AM broadcast stations must produce a directional signal to protect against interfering with other geographically distant stations. They use multiple towers at specific spacing, where the signals are sent slightly out of phase to achieve directivity. Some stations must also be directional even during the day to protect other nearby stations. The phasing is done at the base of each antenna in a box containing an LC network similar to Figure 8.2. The two devices towards the top of the figure are glass-enclosed vacuum capacitors used to tune the antenna. They can handle very high voltages. In this case, the network is used to tune the antenna to a 50 Ohm impedance match to the transmitter and feedline impedance. Directional arrays must stay within the limitations of propagation specified in microvolts (millivolts) per meter readings at designated geographic locations. The measurements are taken with a portable field strength meter.

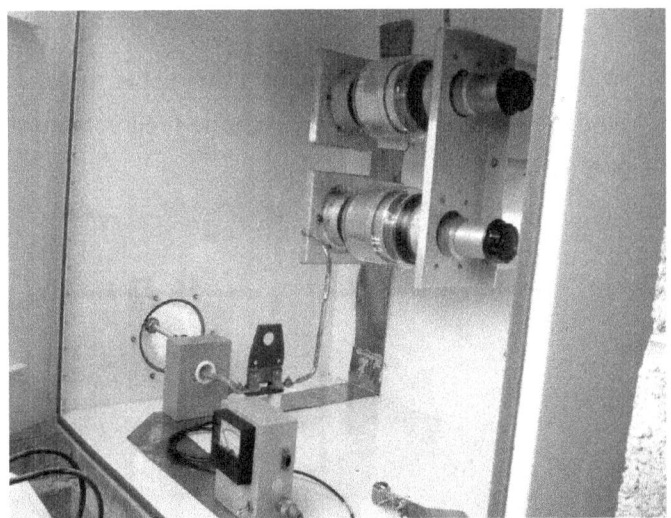

Figure 8.2. Antenna tuning box

Antennas for higher frequencies, such as those used for the FM and TV broadcast band, will usually be insulated from the tower and placed near the top of a grounded tower. The height above ground provides more significant propagation, and the gain is called the Effective Radiated Power (ERP). An FM transmitter may only output 10,000 watts at the final stage but may have an ERP of 50,000 Watts due to the height above average terrain gain. There may also be multiple antenna bays that are somewhat directional in that they direct the signal in what may be considered a "lighthouse" type pattern parallel to the Earth. This increases the gain and produces a higher ERP. Waveguides feed many TV antennas, and the antenna placed high upon the tower uses a slotted design to allow the RF energy to radiate.

Power is best exchanged between the transmitter and receivers if the antenna orientation is the same. It's easy to think about creating waves on the surface of a pond by moving a stick back and forth, where another stick floating in the pond will respond best if it is parallel to the stick generating the waves. Old cars had a small vertical antenna placed on the fender in the years that AM was a more popular broadcast medium. The current windshield style of antennas have a slight reduction in gain for AM, but FM broadcast transmitting antennas are usually circularly polarized to prevent the receiving antenna from being as affected by orientation.

Section 8.2. Horizontal antennas

The most common horizontal antenna is the dipole. It has two sections of equal length, 180 degrees from each other, with the feed point in the center, as illustrated in Figure 8.3.

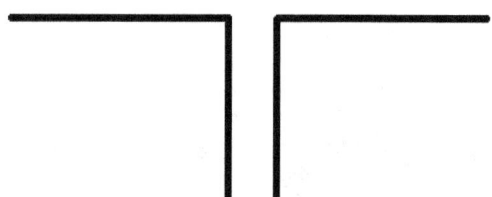

Figure 8.3. Dipole antenna

Each horizontal section of the halfwave dipole is one-quarter wavelength at the desired frequency. The characteristic impedance of a halfwave center-fed dipole is 73 real and 42 Ohms reactive. Moving the feed point to change the radiation pattern and the characteristic impedance is possible. It is standardized that radio transmitters have an output impedance of 50 Ohms. Radio communications coax is 50 Ohms, and cable television is 75 Ohms. The characteristic impedance of coax is independent of its length and depends upon the physical construction between the center conductor and shield. It is necessary to match the impedance along the transmission path as close as possible. If there is a mismatch in the impedance, a reflection will occur. This is not a problem for a receiver's antenna, but if a transmitting antenna is significantly mismatched, the wave reflection can damage the output section of the transmitter. The amount of mismatch is a ratio called the Voltage Standing Wave Ratio (VSWR). If the transmitter, coax, and antenna are all the same characteristic impedance, the VSWR is 1:1, and the reflections are zero. This provides maximum power transfer, but this is an ideal situation that is seldom obtainable. Usually, an impedance matching transformer or a tunable LC network is required to correct mismatches. This is also true for vertical antennas. Horizontal antennas also require a balun (balanced to unbalanced). It isolates the elements from ground and is placed in the path between the radio and antenna.

Horizontal antennas can also be directional with a much simpler construction than the vertical arrays mentioned in the last section. A popular style of horizontal directional antenna called a Yagi is drawn in Figure 8.4.

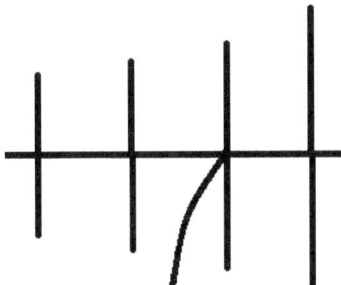

Figure 8.4. Yagi antenna

The curved line in the drawing represents the coax or open-wire feed line connected to the driven element. The larger element(s) behind the driven element reflects the signal towards the front because of its size and spacing. The smaller elements are called directors and focus the RF energy in the direction that the antenna is pointed. The longest piece that holds the elements is called a boom and is electrically isolated from the elements and tower. Yagis are usually orientated horizontally but can also be placed vertically. The orientation of the transmitting antenna should match the orientation of the receiving antennas. Increasing the number of reflectors and directors will increase antenna gain.

The antenna pictured in Figure 8.5 is a portable Yagi used by a Ham radio operator to make a contact with a station across the country through an Amateur Radio satellite.

Figure 8.5. Portable Yagi antenna

The portable antenna is handheld and pointed in the direction of the satellite. A 5 Watt handheld radio is connected through a small length of coax. The large number of elements provides gain, and the RF signal is line-of-sight, so the attenuation is minimal and offers a good connection on the UHF band. Most commercial satellites are much higher in frequency and require a parabolic dish antenna like the one pictured in Figure 8.6.

Figure 8.6. Parabolic dish antenna

The curvature of the dish focuses the satellite's signal at the feed horn, which contains a Low Noise Amplifier (LNA) that boosts the small received signal before sending it through the coax to the television converter box. (LNAs are more recently called LNBs, which stands for Low Noise Block.) The first RF amplifier is placed directly at the antenna to boost the signal at its strongest point in the chain before it encounters attenuation and noise from traversing the coax to the decoder/converter box.

The majority of TV viewers now either get their programs from streaming services, satellite, or through 75 Ohm cable TV, but it is possible to receive free local TV with a small antenna, as pictured in Figure 8.7.

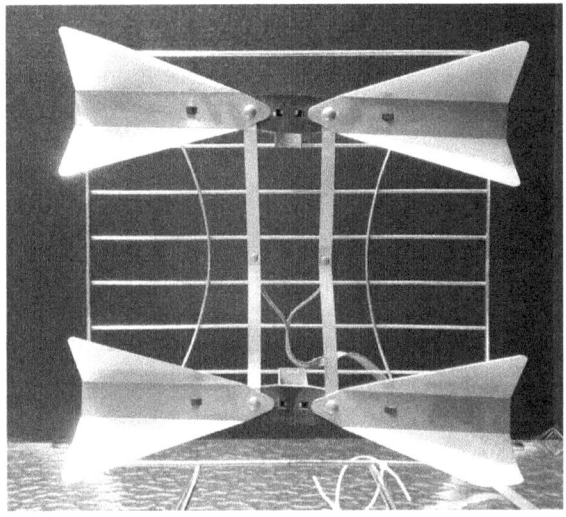

Figure 8.7. Bowtie tv antenna

This old-style tv antenna has a floating grate reflecting the signal to the bowties. Each top and bottom bowtie section connects to one side of the coax. The antenna is directional and sized for UHF reception. The fed line is called ribbon cable, and unlike coax, its impedance is 300 Ohms, so a matching network must be used to connect it to a 75 Ohm TV connection.

There are various styles of antennas, and some Ham Radio hobbyists spend a great deal of time designing and testing new and modified versions for use across the RF spectrum. Electronic communications began with the need for humans to pass messages over long distances. Many technically minded people still find it a fun and exciting endeavor.

Chapter Eight Summary

The physical orientation of antennas are mainly vertical or horizontal, and signals are best transferred if both the transmitting and receiving antennas match. Radio broadcasting stations use vertical antennas, while horizontal antennas are usually used for low through high frequencies communications. Wavelength can be found from frequency, and antenna size should match. The characteristic impedance of a radio, feed line, and antenna should match for maximum power transfer. This is especially important when transmitting, or a high Standing Wave Ratio (SWR) will result, which can damage the power amplifier of the equipment. Yagi and dish antennas are used at higher frequencies. Yagi antennas have reflector and director elements, and dish antennas use a parabolic reflector to concentrate weak superhigh frequencies into a feedhorn containing a low noise amplifier.

Chapter Eight Questions

1. Find the size in feet for a quarter wavelength antenna at 100 MHz.
2. How can vertical antennas be made to have a directional radiation pattern?
3. Why are radials used with vertical antennas?
4. Which broadcast antenna tower is more likely to be electrically floating, an AM or FM station?
5. How can the characteristic impedance of a dipole antenna be changed?
6. What is the characteristic impedance for radio communications coax?
7. What is the characteristic impedance for cable TV coax?
8. What is the purpose of the driven element on a Yagi antenna?
9. Is a director shorter or longer than a reflector on a Yagi antenna?
10. Is communications electronics fun? (The correct answer is yes.)

Appendix

Section A.1. Transducers

A *transducer* is a device that transforms energy from one form to another. Similarly to the way that the ear works, a microphone responds to vibrations in air pressure. A microphone diaphragm is moved by the changing air pressure and converts the *acoustic energy* through mechanical energy into electric energy signals. There are many different conversion mechanisms employed in microphones. Old-style microphones used carbon particles that varied in resistance as the diaphragm either compressed or expanded their density. Some newer technologies use the compression and expansion of a crystal element to generate a voltage. Other methods include varying capacitance, inductance, or magnetic field. Once the transducer generates a voltage, electronic circuits can then amplify it.

When we use our voice, electrical impulses cause muscles to constrict and make the vocal cords vibrate. This transducer is an example that converts electrical energy into mechanical energy to produce an acoustic wave. A speaker or headphone is a very similar type of transducer. The speaker converts electrical energy into mechanical energy to cause the air to acoustically vibrate. Most speakers are electromechanical and use a movable coil of wire, called the voice coil, fastened to a heavy paper or plastic diaphragm. A current passes through the coiled wire to create a magnetic field. The magnetic field of the voice coil interacts with the magnetic field of a stationary magnet causing movement of the diaphragm. The polarity of the voltage across the coil, and the resultant direction of current flow, determine the direction of movement of the diaphragm. Most large speakers, such as woofers, work in this way. Tweeters and other small speakers tend to use a *crystal*. Just as the crystal microphone generates a voltage as the element is distorted by being bent back and forth by the moving diaphragm, a voltage signal sent to a crystal causes the crystal to move. A crystal will bend either back or forth, depending on the polarity of the voltage across it. Microphones and speakers can use the same concepts of operation in opposite ways. This is an example of *reciprocity*.

When we discuss electrical signals used to produce a sound, or generated from a received sound, we are dealing with signals representing waves. The waves are sinusoidal in nature. Remember that a sine wave has a positive and negative alternation. The characteristics of the sinusoidal sound waves are similar to the AC electric waves that we discussed earlier in the text. Electromagnetic waves are similar but contain an electric and magnetic component without any acoustic vibration of the air. The waves are analog because they do not have any *discontinuity*. When we studied digital electronics, we learned that transistors could be used to switch on-and-off to correlate to ones and zeros. In analog electronics, the transistor is operated between the completely on or off state, and used to amplify the analog signals that gradually change between high and low levels.

A.2. Amplification

Most microphones and speaker systems are passive, which means they do not need electric power in addition to the signal for them to operate. An amplifier is active since a transistor needs applied power to be in the state between being completely on or off. We use the term *cut-off* when a transistor is completely off, and *saturation* describes a transistor completely in the on state. In digital circuits, the transistor is operated in these two extremes. The voltage across a *cut-off transistor* is maximum, whereas the current is minimum. The voltage across a *saturated transistor* is minimum, whereas the current is maximum, and vice versa. On a graph of current vs. voltage, a line drawn as in Figure 13.2, between the two extremes is called a *load line*.

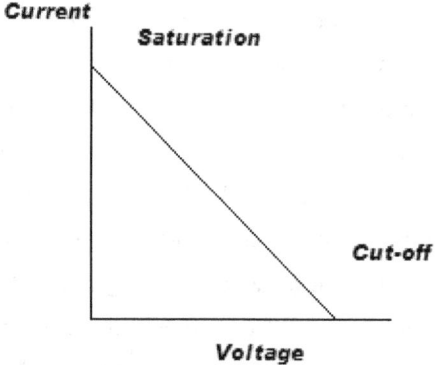

Figure 13.2. Load line collector to emitter

There are different classes of amplifier operation relating to where conduction takes place on the load line without an input signal. That point is called the Q point and comes from the word *quiescent*. For the best fidelity, an amplifier's Q point is located in the middle of the load line. That location provides little chance of distortion from the signal driving the amplifier to either hit saturation or cut-off, causing a distortion condition called *clipping*. (Clipping happens when the amplifier reaches its limit.) This class of amplifier is considered *class-A*. This class has the best fidelity but poor power efficiency because it is conducting 100% of the time. The other classes conserve power but have a trade-off with the fidelity.

More than one transistor is usually needed to amplify small signals. Each transistor section will make the signal voltage greater by a certain amount called *gain*. We use the letter A to represent the level of amplification. Each section is called a stage, and the total voltage gain is the multiplication of each of the stages' gains. This is how many times the input signal voltage is multiplied by all stages of amplification, as shown in the figure.

In our example, A = (A1)(A2)(A3) = (10)(10)(10) = 1000. The input signal

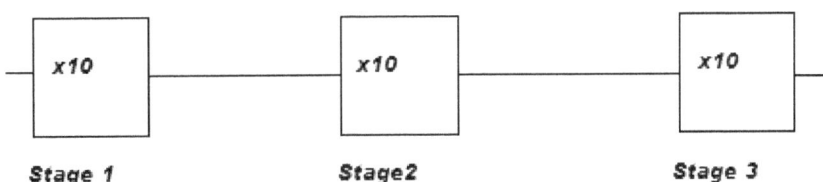

coming from the left is multiplied 1000 times as it goes through all of the stages and leaves on the right.

Problem

Find the final output voltage in the above figure if the input level is 18 mV.

Solution

The unit mV in the problem is in engineering notation, and the *m* is for milli.

A_t = (A1)(A2)(A3) = (10)(10)(10) = 1000

The output = (Input)(A_t} = (18 x 10^{-3} volts)(1 x 10^3) = 18 volts.

Notice that there is no unit associated with gain (A). Many times the gain of an amplifier is expressed in *decibels* (dB). The individual gains of each stage in dB are added together to find the total dB gain. For voltage, the following formula is used to convert to dB:

$$dB = 20 \log A$$

$$\text{Where A is } \frac{Output}{Input}$$

Problem

In the previous problem, express the individual gains and the total gain of all the stages using decibels.

Solution

For each stage:

$$dB = 20 \log A$$
$$dB = (20) \log 10$$
$$dB = (20)(1) = 20$$

For the total gain expressed in dB:

$$\text{Total dB} = 20 + 20 + 20 = 60 \text{ dB}$$

We could verify our answer by using the answer that we get for A_t directly.

$$dB = 20 \log A_t$$
$$dB = 20 \log 1000$$
$$dB = (20)(3) = 60$$

We have mainly been discussing voltage amplifiers. In order to drive speakers from a stereo system, the voltage amplifier stages must feed a power amplifier section. Class-A operation would waste a lot of power. Most power amplifiers are designed to run so that there is little current draw without an input signal. It is called push-pull operation and uses two transistors. One conducts during the positive going alternation of the input signal, while the other conducts during the negative half. The power amplifier Q point is very low on the load line diagram that we looked at previously, and close to cut-off. In problems that you wish to express power gain or loss in dB, the formula is slightly different:

$$dB = 10 \log A$$

There are a few quick rules of thumb for power expressed in dB. 3 dB is double power, and 10 dB is equal to a gain of 10 times. We added a chart for common dB relationships in Section A.4.

Sometimes there are circuits where you may want to reduce voltage or power. Reducing a signal is called *attenuation*. If a stage has less signal level coming out than is going in, the gain A and the dB will have a negative sign associated with it. We have a chart of common dB values in appendix A4.

Section A.3. Operational Amplifiers

With the rapid adoption of the IC, very few discrete transistor audio amplifiers are produced today. Entire analog circuits are contained in *linear* ICs. The word linear describes the operation. In linear amplification, the output is raised from the input signal in a direct relationship that is constant. If the input signal were plotted against the output, the graph would form a straight line. A versatile IC voltage amplifier called an Op-Amp is very popular because it is inexpensive, easy to use, and requires few external components. One drawback is that they sometimes require a split voltage supply to power the device. It must have equal but opposite voltages, one positive and one negative. There are, however, a few device types that only need a single supply. The Op-amp can be designed to either *invert* or *non-invert* the signal. Usually, nobody would ever know if the input signal was inverted as an output since it would sound the same. We will describe the inverting Op-amp operation shown in Figure 13.4 since it is easier to understand.

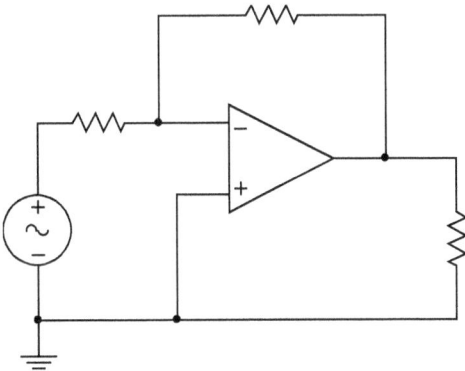

Figure 13.4. Amplifier Circuit

The diagram of an Op-amp is a triangle that points to the output. In our drawing a sine wave signal source on the left is connected through an input resistor to the negative sign on the Op-amp. The negative sign identifies the inverting input. The other resistor that is connected to that point is called the *feedback* resistor. It takes part of the inverted output signal and sends it back to the input. Since the output is inverted, it has a signal that is the opposite. The negative feedback subtracts from the input level. The gain of the stage is determined through the selection of the ratio of the feedback resistor to the input resistor. The other resistor on the right of Figure 13.4 develops the output voltage. The voltage gain of an inverting Op-amp can be found with the following formula:

$$A = \frac{R_f}{R_i}$$

Problem

Determine the gain of an Op-amp circuit if the feedback resistor is 10 kΩ and the input resistor is 1 KΩ.

Solution

Remember the symbol Ω is the Greek letter Omega and is the symbol for resistance. Now, using our formula:

$$A = \frac{R_f}{R_i}$$

$$A = \frac{10,000}{1000} = 10$$

The output voltage would be 10 times larger than the input.

Op-amps also need to have two IC pins connected to power the IC. They are not shown in our drawing. Using Op-amps in the design of amplifiers has many advantages. They have a high input impedance. You may remember that impedance is very similar to resistance. With a high input impedance, the stage in front of the Op-amp would not be affected by the connection. Op-amps also have a low output impedance. That is good because there is very little opposition in the Op-amps output to supplying current to the following stage, should it require it. Op-amps are designed to be voltage amplifiers. There are other ICs available that are power amplifiers. An Op-amp that is in wide use has the part number 741, and a common IC power amp is part number 386. For a linear IC, the prefix is a group of letters that identify the manufacturer. An LM741 is a 741 Op-amp made by National Semiconductor. There is a different system for digital IC part numbers. For digital

ICs, the base number is preceded by letters that identify the type of technology, and the part number begins with 74 to identify the IC as commercial production, or 54 to signify that it meets military specifications. For a 74LS08, the 08 is the base number for an AND gate, LS is low power Schottky technology, and 74 identifies it as suitable for commercial use. It is very easy to search for an IC part number on the World Wide Web, and download a manufacturer's datasheet at no charge. The datasheet will list all of the specifications for the device, as well as a pin-out diagram to identify the function of each connection pin.

The pin-out assignments for *digital* ICs are very standardized. On almost all digital Dual Inline Package (DIP) ICs, the ground connection is the lower rightmost pin, and the power connection is the upper leftmost. The reason for this standardization was because, in the past, many digital ICs needed to be connected together on a circuit board. This IC pin-out orientation made it easier to lay out the ground and power busses on the board. Analog ICs do not adhere to this convention, and each device will have different pin assignments. The numbering system, however, is the same for all ICs. An orientation mark, usually a semicircle or circle, is placed on the left, as appears in Figure A.3, and the lower leftmost pin is number 1. The numbers go along the pins counter-clockwise, with the highest number assigned to the upper leftmost pin.

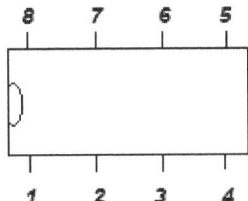

Figure A.3. Pin numbering of an IC

A.4. Decibel information

The decibel (dB) has been around since the beginnings of landline telephone service. It comes from a unit named after Alexander Graham Bell called the Bell which is a log of a ratio. The decibel is more convenient and equal to $1/10^{th}$ of a Bell. Audio and RF power are not linear but logarithmic and the reason dBs are used quite a bit in electronic communications. The following table lists some of the more common dB conversions for power and voltage:

Decibel conversions

dB	Power	Voltage
+3	x 2	--
-3	1/2	--
+6	x 4	x 2
-6	1/4	1/2
+10	x 10	--
-10	1/10	--
+20	x 100	x 10
-20	1/100	1/10

A.5. Parts list

Arduino Uno with USB cable

Small breadboard

Small speaker (3 inch)

Breadboard hook-up wires

40 cm 22-gague wire

Electret condenser microphone

NE555 timer IC

LM386 power amp IC

2N3904 NPN transistor

HC-05 Bluetooth module

NTC MF52-103 10k Thermistor

433 MHz transmitter and receiver

IR diode

IR receiver

(5) LEDs

Capacitors in uF: 0.01, 0.1, 1, (2) 10, 220

Resistors: (2) 100, 120, 200, (4) 220, 680, 1 k, 1.5 k, 2.2 k, 4.7 k, 10 k, 22 k

www.ingramcontent.com/pod-product-compliance
Lightning Source LLC
Chambersburg PA
CBHW060418220526
45465CB00008B/2928